T0318864

An Introduction to Organic Lasers

Advanced Lasers Set

coordinated by
Pierre-Noël Favennec, Frédérique de Fornel
and Pascal Besnard

An Introduction to Organic Lasers

Azzedine Boudrioua
Mahmoud Chakaroun
Alexis Fischer

ELSEVIER

First published 2017 in Great Britain and the United States by ISTE Press Ltd and Elsevier Ltd

ISTE Press Ltd
27-37 St George's Road
London SW19 4EU
UK

www.iste.co.uk

Elsevier Ltd
The Boulevard, Langford Lane
Kidlington, Oxford, OX5 1GB
UK

www.elsevier.com

Notices

Knowledge and best practice in this field are constantly changing. As new research and experience broaden our understanding, changes in research methods, professional practices, or medical treatment may become necessary.

Practitioners and researchers must always rely on their own experience and knowledge in evaluating and using any information, methods, compounds, or experiments described herein. In using such information or methods they should be mindful of their own safety and the safety of others, including parties for whom they have a professional responsibility.

To the fullest extent of the law, neither the Publisher nor the authors, contributors, or editors, assume any liability for any injury and/or damage to persons or property as a matter of products liability, negligence or otherwise, or from any use or operation of any methods, products, instructions, or ideas contained in the material herein.

For information on all our publications visit our website at http://store.elsevier.com/

British Library Cataloguing-in-Publication Data
A CIP record for this book is available from the British Library
Library of Congress Cataloging in Publication Data
A catalog record for this book is available from the Library of Congress
ISBN 978-1-78548-158-1

Printed and bound in the UK and US

Contents

Foreword

Organic lasers represent the start of a promising new avenue for laser technologies. They use a hugely abundant source material – carbon. Is it possible that they will one day come to replace semiconductor lasers, and what sort of performances will they deliver? The organic laser will indubitably be an innovative component: it will be a low-cost laser, built with materials from an unlimited source, unlike devices using gallium or indium, of which there are very scant mineral reserves. But what will the structures be? What levels of performance can we expect? The technologies are complex and, as yet, imperfectly understood, and the performances of those organic lasers which have been created are still a long way behind those obtained with conventional semiconductor lasers.

Given the relative rarity of their base materials, semiconductor lasers cannot possibly cater for society's massive need for lasers, and in particular, for domestic use lasers which do not require extraordinary performances.

Organic materials, all having carbon as a base element, may, depending on their composition and treatment, act as insulators, conductors or semiconductors. With such properties, we can envisage being able to technologically create structures of electroluminescent (light-emitting) diodes and laser diodes. However, this will require a good knowledge of organic materials, in terms of their electrical, optical, thermal, mechanical (etc.) properties, and excellent stability of all these properties when processed technologically: thin films, engraving, localized doping, stacking of layers, electrical contacts, and so on.

This book describes the whole process of transforming organic materials into laser diodes: the materials and the technological problems they pose are examined, followed by the fabrication of OLEDs and their functioning principles, and finally laser diodes. The fundamental principles for these organic lasers are well established and familiar. However, a great deal of technological progress is still needed before we can create components whose performances are acceptable and stable, and in particular, which are highly reliable.

One of the merits of this excellent work is its projection into the future of laser diode components into nanophotonics through nanolasers. This vision for nanophotonics lies at the crossroads between two branches of physics: physics of lasers and physics of surface plasmons.

Organic lasers are undeniably a category of the lasers of the future, but for low-cost, widely-available devices, unlike other types such as those based on III-V semiconductors and innovative, sophisticated structures. At present, they are merely laboratory objects, but we can see that they will very soon experience mass expansion. Their widespread dissemination appears to be a foregone conclusion.

I would recommend this book for everyone in laboratories working on organic material problems, and all those thinking about tomorrow's technologies, to create ever more useful components for multiple applications. This book could also constitute a teaching resource for Masters or degree-level courses on physics of lasers.

Pierre-Noël FAVENNEC

Acknowledgements

This book is a collective work, essentially based on the research and teaching at the Paris 13 University of the *Photonique Organique et Nanostructures* (Organic Photonics and Nanostructures) team in the laser physics lab. Notably, it uses the various thesis manuscripts and reports supervised by the authors. Indeed, for the purpose of clarity and efficiency, this book was prepared based on texts collected from the different doctoral thesis manuscripts written under our supervision in recent years.

Thus, we offer heartfelt thanks to the PhD students who allowed us to use passages or figures from their manuscripts. Every one of them has, in one way or another, contributed to the writing of this book. Particular thanks go to Samira Khadir, Lei Zeng, François Gourdon and Anthony Coens.

A very big thanks, too, to Pierre-Noël Favennec, who came to us with the idea of this project, and has very patiently followed its development.

We also wish to thank ISTE for agreeing to publish this book, and in particular for the patience and close monitoring of the progress of this project.

Azzedine BOUDRIOUA
Mahmoud CHAKAROUN
Alexis FISCHER

Introduction

Organic lasers have punctuated the history of lasers almost from the very beginning. They truly began being developed, though, in 1966, with the almost simultaneous demonstration, by multiple teams in the USA and Germany, of dye lasers.

More recently, a large variety of solid-state lasers using different types of cavities or nano-cavities with optical pumping have been demonstrated. In parallel, the advances achieved in the field of organic light-emitting diodes (OLEDs) allow the electrical excitation of organic materials at high current density. Today, the scientific issues involved in these new and future organic lasers are a logical continuation of this scientific and technological story. Work is currently being done to create the first organic laser diode based on organic semiconductors by electrical excitation. Indeed, the substitution of an optical excitation with an electrical current in an organic gain medium requires a set of modifications, from the transformation of the laser cavities to the design of new organic materials, or hybrid organic–inorganic hybrids. A new multidisciplinary approach is needed to deal with all aspects of this evolution of lasers.

The spirit of this book, therefore, is to present the elements of physics, materials and technologies that account for the progress made, enabling us to envisage the logical evolutions toward organic laser diodes.

In order to be able to quantitatively present this scientific and technological challenge, and the progress still to be made to meet that challenge, and also to gain a clearer idea of the feasibility, the pedagogical approach is divided as follows: firstly, we consider OLEDs, which are the closest scientific objects to organic laser diodes. Secondly, we present optically pumped organic lasers, and notably solid-state ones, whose lasing thresholds are identified. Finally, we examine the possible avenues of evolution, and focus on the domain of organic nano-photonics in the process of development. This research could also provide new solutions to advance towards organic laser diodes.

Thus, this book is structured as follows:

– Chapter 1 is devoted to a general recap about organic materials, and more specifically, organic semiconductors, which constitute the cornerstone of organic lasers;

– Chapter 2 is dedicated to OLEDs, which are used as a way in, to illustrate the principles of electrical pumping in organic semiconductors;

– Chapter 3 is given over to solid-state organic lasers;

– finally, Chapter 4 presents the prospects opened up by organic nano-photonics and plasmonics to develop new organic laser architectures.

The book ends with a general conclusion, which also discusses some avenues for future development.

Organic Semiconductors

The aim of this chapter is to recap the elementary properties of organic materials, which are the cornerstone of organic optoelectronics. In order to understand this new class of materials, it is important to begin by looking at organic chemistry basics. Based on concepts taken from organic chemistry, it is possible to underline the physical, electronic, and also optical properties, which are crucial for understanding how organic optoelectronic devices, such as lasers or organic light-emitting diodes (OLEDs), work.

The first section recaps the quantum model of the atom, enabling us to present both the electronic properties with their corollary in terms of conduction and transport, and the optical properties and the molecular structure of the materials used in laser devices and OLEDs. The chapter closes with a presentation of a number of families of molecules which are particularly important in the history of organic lasers. These reminders are intended primarily for students newly come to the discipline, but they also represent the fundamentals that must be known in order to understand organic emitter components.

1.1. Recap on organic chemistry

Although Lewis' model is able to explain a certain number of properties of organic molecules, it is not sufficiently complete to explain, in particular, the limitations in light emission yield from certain organic molecules. For this reason, it is crucial to use the quantum model of the atom, based on Schrödinger's equation, to describe the evolution of the electrons around the nucleus of the atom. Atoms, composed of a nucleus including protons and neutrons, and electrons orbiting the nucleus, are indicated in the periodic table of the elements by a symbol (X), their atomic number (Z) and their mass number (A):

– X: symbol of the element. X = C for carbon, H for hydrogen, O for oxygen and N for nitrogen;

– A: the mass number indicates the number of nucleons in the nucleus – i.e. the number of neutrons plus the number of protons (charges +e = +1.6×10^{-19} C);

– Z: the atomic number indicates the number of protons present in the nucleus. As atoms are electrically neutral, this is also the number of electrons.

Thus, carbon $C_{Z=6}^{A=12}$ has 12 nucleons, including six protons, and therefore six electrons; hydrogen H_1^1 has one proton and one electron, and oxygen O_8^{16} has a mass number of 15.99, representing 16 nucleons, including eight protons, and therefore eight electrons.

1.2. Quantum model of the atom

The number of electrons and the electron configurations of atoms are essential information for understanding, firstly, the possibilities of arrangement of the atoms to form organic molecules, and secondly, the properties of those molecules. By solving Schrödinger's equation, we find the electron states using a wave function involving 4 parameters, which are the quantum numbers n, m, l and s, that enable us to define the electron layers, the electron sublayers, quantum cases and electron spin:

– n is the principal quantum number: it is a strictly positive natural number (n > 0) which defines the number of an electron layer. The set of electrons of an atom which have the same number n constitutes an electron layer. A layer n will contain at most $2n^2$ electrons. Those layers are named K, L, M and N for n = 1, 2, 3, 4, respectively;

– l is the secondary or azimuthal quantum number: this too is an integer which depends on n, because it is $0 \leq l \leq (n–1)$. Electrons having the same number l constitute an electron sublayer.

If n = 1, then l can only take the value of l = 0.

If n = 2, then l can take the values l = 0, l = 1.

If $n = 3$, then l can take the values $l = 0$, $l = 1$, $l = 2$.

The different values of $l = 1, 2, 3, \ldots$ correspond respectively to sublayers called s, p, d and f;

$-m$ is the magnetic quantum number such that $-l < m < +l$. Within a sublayer, the magnetic quantum number defines the number of quantum states. The electrons in the same sublayer, having the same magnetic quantum number m, belong to the same quantum state. There are at most two electrons per quantum state.

For example, for the layer $n = 1$, sublayer s ($l = 0$), there is only one quantum state ($m = 0$), and for the layer $n = 2$, sublayer p ($l = 1$), there are three quantum states ($m = -1$, $m = 0$, $m = +1$):

$-s$ is the spin quantum number. Electrons can take one of two spin values: $s = \frac{1}{2}$ and $s = -\frac{1}{2}$. They are represented by arrows – an upward arrow for a spin of $+\frac{1}{2}$ and a downward one for a spin of $-\frac{1}{2}$. There can only be at most two arrows in each quantum box. When two arrows do occupy the same quantum state, they must be of the opposite direction: ↑↓.

The table below summarizes electron configurations.

n (layers)	l (sublayers)	m (quantum cases)	s (spin)
n=1 (K)	l=0 (1s)	0	+1/2 (↑) and −1/2(↓)
n=2 (L)	l=0 (2s)	0	+1/2 (↑) and −1/2(↓)
	l=1 (2p)	−1	+1/2 (↑) and −1/2(↓)
		0	+1/2 (↑) and −1/2(↓)
		1	+1/2 (↑) and −1/2(↓)
n=3 (M)	l=0 (3s)	0	+1/2 (↑) and −1/2(↓)
	l=1 (3p)	−1	+1/2 (↑) and −1/2(↓)
		0	+1/2 (↑) and −1/2(↓)
		1	+1/2 (↑) and −1/2(↓)
	l=2 (3d)	−2	+1/2 (↑) and −1/2(↓)
		−1	+1/2 (↑) and −1/2(↓)
		0	+1/2 (↑) and −1/2(↓)
		+1	+1/2 (↑) and −1/2(↓)
		+2	+1/2 (↑) and −1/2(↓)

Table 1.1. *Recap of possible electron configurations*

1.2.1. *Electronic structure of atoms*

The electronic structure (or electron configuration) of an atom defines the way in which the Z electrons in the atom are divided between the layers, sublayers, quantum states and spins. This electron configuration explains, and enables us to predict, an atom's electronic and optical behavior. The Pauli exclusion principle, Hund's rule and Klechkowski's rule are the three rules governing the filling of orbitals, which specify how the different quantum states described above can be occupied by the different electrons of the same atom.

Klechkowski's rule dictates the order in which the layers and sublayers are filled: the sublayers are filled so that the sum $n + l$ increases. First, the cases where the $n + l = 1$ are filled, followed by that where $n + l = 2$, then $n + l = 3$, etc.

In case of equality, the sublayer with the smallest value of n is filled first.

Thus, the sublayer 2p, corresponding to $(n = 2, l = 1,$ so $n + l = 3)$ is filled after the sublayer 2s corresponding to $(n = 2, l = 0,$ so $n + l = 2)$, and before the sublayer 3s $(n + l = 3)$.

The sublayer 4s $(n = 4, l = 0, n + l = 4)$ is filled before 3d $(n = 3, l = 2, n + l = 5)$. The order of the cases is illustrated in Figure 1.1.

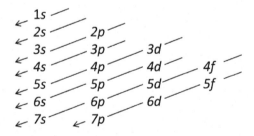

Figure 1.1. *Klechowski's rule specifies the order in which quantum cases are filled with electrons*

Hund's rule: within a sublayer, the electrons first arrange themselves in a configuration of one electron per quantum case. They arrange in pairs, matching up with one another to form doublets when they are more numerous than the cases in the sublayer in question.

The Pauli exclusion principle specifies that two electrons cannot occupy the same quantum state in the same system (atom). This means that the electrons in the same atom have their four quantum numbers n, m, l and s which are all different to one another.

1.2.1.1. *The atomic orbitals s and p*

By solving Schrödinger's equation, we are able to describe how the electrons occupy the regions around the nucleus of the atom in a probabilistic fashion. The volumes of space where the probability of finding electrons reaches 95% for the different values of n, l and m are called orbitals. The electron state corresponds to a wave function which depends on n, l and m. The energy of the electron depends on n and l.

1.2.1.2. *The s orbitals*

The s orbitals depend solely on the radius r (distance between the center of the atom and the position in question), and thus have spherical symmetry. The shapes with 95% probability of presence of the different sublayers s depend on the radius r:

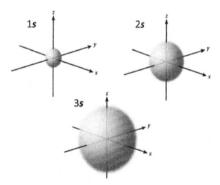

Figure 1.2. *The s orbitals exhibit spherical symmetry which delimits the zone of space in which the probability of finding an electron is 95%*

1.2.1.3. *The p orbitals*

The p_x, p_y, and p_z orbitals, respectively, show symmetry along the x, y and z axes.

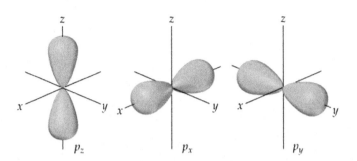

Figure 1.3. *The p_x, p_y and p_z orbitals respectively exhibit cylindrical symmetry along the x, y and z axes*

1.2.2. *Molecular orbitals*

1.2.2.1. *Core layer and valence layer*

"Core electrons" are those in the core layer – i.e. a layer which is completely filled. All the quantum cases in a layer contain two electrons with each spin. The elements in the periodic table having only core electrons exhibit great stability and are therefore rather unreactive.

The "valence layer" is the peripheral layer of an atom – i.e. the electron layer, partially or completely filled, which corresponds to the highest principal quantum number n. We use the term "valence electrons" to speak of those belonging to the valence layer. The presence or absence of valence electrons in the quantum cases of the valence layer has a significant influence on the reactivity of elements, and consequently on their chemical and electrical properties. For atoms whose valence layer is not filled, by interacting or associating with other atoms, pooling electrons, they are able to reach more stable, lower-energy electron configurations. By the pooling of electrons, certain atoms can be combined more easily with others to complete their valence layers.

1.2.2.2. *Bonds between atoms*

Atoms can bond with one another in two different ways:

– a covalent bond: in this case, each of the atoms shares one of the electrons in its valence layer, occupying a single quantum case. The two electrons (doublet) pair up to form a covalent bond common to the two atoms:

$$A. \rightarrow \leftarrow .B \Rightarrow A:B \text{ or } A-B \qquad [1.1]$$

– co-ordinate bond: in this case, one of the atoms donates a doublet of electrons from its valence layer to the accepting atom, where they are placed in an empty case in its valence layer: $A :\rightarrow B \Rightarrow A:B$ or $A-B$.

Note that, atoms may share several doublets (two or three), and thus create double or triple bonds.

1.2.2.3. Orbital overlap

When atoms share electrons and come together to form molecules, the orbitals of each atom interact, overlap and form molecular orbitals. This overlap must be seen as the result of the linear combination of atomic orbitals. Thus, the wave function Ψ of a molecule made up of two atoms A and B is a linear combination of the wave functions ψ_A and ψ_B. Two types of combinations, and thus two types of overlap, are possible depending on the sign of the wave functions: $(\psi_A+\psi_B)$ or $(\psi_A-\psi_B)$.

– when the wave functions of the two atoms A and B overlap with the same sign $(\psi_A+\psi_B)$, the overlap between orbitals is constructive, and the probability of finding an electron in the overlap zone is increased (see Figure 1.4a). When two electrons occupy such a molecular orbital, they help bond the two atoms together, and hence, are known as a bonding orbital. A bonding molecular orbital corresponds to a lower level of energy than that of each of the atoms taken in isolation;

– when the wave functions of atoms A and B overlap with opposite signs $(\psi_A-\psi_B)$, the overlap between orbitals is destructive, and the probability of finding an electron in that region of space is lesser (see Figure 1.4b). They contribute to a destabilization of the chemical bond between the atoms, and are characterized by a higher energy than that of each of the corresponding atomic orbitals taken separately, so they are known as antibonding molecular orbitals. They are written with an asterisk, *;

– when wave functions of different symmetries overlap, that overlap may be slight or even null (Figure 1.4c). Such is that case, for example, with the overlap of atomic orbitals s and p. In this case, we speak of nonbonding molecular orbitals. The energy of nonbonding molecular orbitals are the same as those of the atomic orbitals making them up.

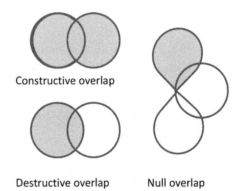

Constructive overlap

Destructive overlap   Null overlap

Figure 1.4. *Different types of overlaps between orbitals*

Bonding molecular orbitals can be obtained by three types of overlap:

– constructive overlap of *s* orbitals, as illustrated in Figure 1.5:

Figure 1.5. *Overlap of s orbitals*

– overlap of p_z orbitals, as illustrated by Figure 1.6. Remember that, by convention, the z axis is the axis passing through the center of the atoms:

Figure 1.6. *Overlap of p_z orbitals*

– overlap of p_x and p_y orbitals, exhibiting a smaller overlap, because it occurs on an axis different to the z axis:

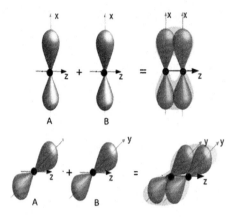

Figure 1.7. *Overlap of p_x or p_y orbitals*

Anti-bonding molecular orbitals can also be obtained by three types of destructive overlaps:

– destructive overlap of orbitals s;

– destructive overlap of orbitals p_z;

– destructive overlap of orbitals p_x, p_y.

Non-bonding overlaps are obtained in the case of null overlap between an s orbital and a p_x or a p_y orbital (see Figure 1.8):

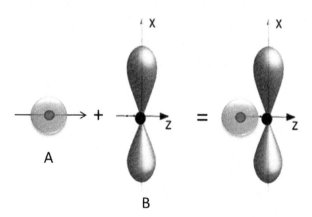

Figure 1.8. *Null overlap between an orbital s and an orbital p_x*

1.3. Sigma (σ) and pi (π) bonds

The different types of overlap of valence electrons from an atom associated with another atom define two types of bond: σ bonds and π bonds:

– sigma (σ) bonds: σ bonds are overlaps of atomic orbitals producing bonding molecular orbitals whose axis of symmetry is the axis linking the two atoms (conventionally the z axis). Sigma bonds are the strongest kind of covalent bonds, because the overlap along the axis passing through the center of the two atoms is greater than the lateral overlap. It may be:

    - constructive overlap of atomic orbitals s (Figure 1.5),

    - constructive overlap of p_z orbitals (Figure 1.6),

    - antibonding bonds which correspond to the destructive overlap of s or p_z orbitals are written as $s*$ bonds and are called conjugate sigma bonds;

– π bonds: π bonds correspond to molecular orbitals with a plane of symmetry passing through the z axis linking two atoms and whose axis of symmetry is perpendicular to the axis linking the two atoms. In this case, we have the overlap of p_x or p_y orbitals (see Figure 1.7). Conjugate π bonds (π*) correspond to the molecular orbitals obtained by destructive overlaps of antibonding atomic orbitals.

1.4. Example of molecular orbitals for simple molecules

1.4.1. *Example of the dihydrogen molecule*

Figure 1.9 shows the energy associated with the molecular orbitals and also illustrates the principle of construction of those molecular orbitals in the simplest possible case: that of the dihydrogen molecule H_2. On the left and the right, the circles represent the atomic orbitals of separate hydrogen atoms. Their electrons are in the base state characterized by the quantum states $n = 1$, $l = 0$, $m = 0$ and $s = +1/2$, represented by the upward arrow. The corresponding energy level is $1s$. When those two atoms are brought together to form the dihydrogen molecule, the two atomic orbitals overlap and partially merge, forming the molecular orbital with the two possible configurations. These two configurations correspond to the two linear combinations $(\psi_A + \psi_B)$ and $(\psi_A - \psi_B)$, facilitating the association of the wave functions ψ_A and ψ_B of each of the atoms H_A and H_B:

– the bonding state ($\psi_A + \psi_B$) obtained with the two wave functions of the same sign corresponds to a partial but constructive overlap of the orbitals, written as $s_{H\text{-}H}$ in Figure 1.9. The figure clearly shows that the energy level of $s_{H\text{-}H}$ is lower than the level of energy of each of the atoms separately;

– the antibonding state ($\psi_A - \psi_B$) represented by $s^*_{H\text{-}H}$ in Figure 1.9, corresponds to an overlap of the orbitals with wave functions of the opposite sign. The energy is then higher than the energy level of orbitals of each of the hydrogen atoms individually.

Thus, two atoms of hydrogen A and B forming the dihydrogen molecule are held together by a sigma bond which corresponds to the partial overlap of two s orbitals as shown in Figure 1.9.

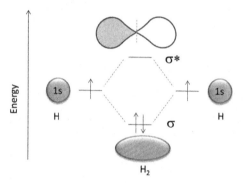

Figure 1.9. *Representation of the formation of the H_2 molecule*

1.4.2. *The case of carbon*

Carbon is central to organic materials, and also exhibits peculiar qualities in terms of its electronic structure, lending it electrical and optical properties that are important in organic lasers.

Having six electrons, the element carbon has two electrons in the first layer and four electrons in its outer layer $n = 2$. The distribution of the electrons in that layer is therefore, *a priori*, two electrons on the sublayer $2s$ and two electrons on two of the three quantum cases $2p$. However, carbon is capable of forming four bonds with four hydrogen atoms to form the methane molecule (CH_4), illustrated in Figure 1.10.

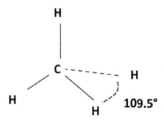

Figure 1.10. *Tetrahedral structure of the methane molecule. The angles formed between the C–H bonds are all identical and equal to 109.05°*

Experience tells us that the structure of methane is tetrahedral, and that carbon–hydrogen bonds (C–H) form identical angles of 109.05°. These bonds have the same energy, which appears inconsistent, because the electrons in the quantum cases 2s and 2p have two different energies.

1.4.3. *Hybridization of the carbon atom*

The theory of hybrid orbitals or hybridization, put forward by Linus Pauling [CAR 12], explains particular geometries of molecules, such as that of methane. Hybridization is the transformation of the atomic orbitals so that they better describe the molecule's properties. Generally, hybridization is the mixing of an atom's atomic orbitals in the same electron layer, to form new orbitals which better account, qualitatively, for the bonds between atoms. Hybrid orbitals are very useful in explaining the shape of molecular orbitals.

To explain this structure, we need two additional concepts:

– the different distribution of the electrons outside of the base state. Indeed, when not in the base state, the electrons in the last layer are liable to be distributed differently. In the case of methane, carbon incompletely, but identically, fills each of the four quantum cases with a single electron (one 2s quantum case and three 2p quantum cases). It should be noted that two different types of orbitals are involved: the 2s and the 2p orbitals;

– modification (hybridization) of the orbitals: the orbitals that are known exactly from solving Schrödinger's equation are those of hydrogen. In the case of heavier atoms such as carbon, oxygen and nitrogen, the orbital of an electron is deformed, because the influence of the effective charge, in

Schrödinger's equation, is that of the nucleus of the atom decreased by the other electrons (attraction by the nucleus and repulsion by the other electrons). The solutions are then new mathematical functions called hybrid orbitals, denoted as sp^3, sp^2 or sp.

1.4.4. The sp^3 hybridization of carbon

The s and p orbitals, instead of remaining distinct, are reorganized to form four identical orbitals, oriented along four axes, forming a regular tetrahedron whose points form an angle of 109.28° with each other.

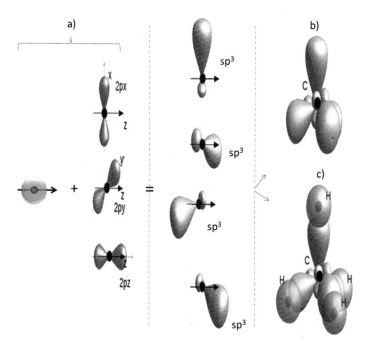

Figure 1.11. Illustration of sp^3 hybrid orbitals: a) Principles of reorganization of the orbitals; b) geometric form of sp^3 hybrid orbitals; c) example of configuration and bonds involved in the case of the methane (CH₄) molecule

In the methane molecule, the four hybridized orbitals of carbon overlap with the s orbitals of each of the four hydrogen atoms to form s bonds orientated along the axis passing through the center of the nuclei of the carbon and hydrogen atoms (see Figure 1.11). Those same orbitals are present in the tetrahedral structure of the carbon crystal (diamond).

1.4.5. Sp^2 hybridization of carbon

In the case of sp^2 hybridization, it is the orbitals $2s$, $2p_x$ and $2p_y$ (Figure 1.12(a)) which are reorganized, rather than remaining separate, to produce three identical orbitals, orientated along three coplanar axes 120° apart. The orbital $2sp_z$, which is not involved in this hybridization, remains perpendicular to the plane of these three coplanar sp^2 orbitals.

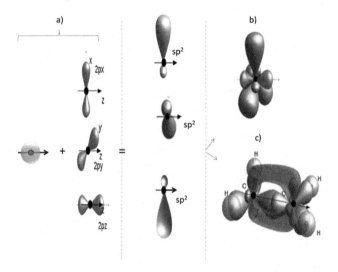

Figure 1.12. *Illustration of sp^2 hybrid orbitals: a) Principle of reorganization of the orbitals; b) geometric orientation of the different orbitals; c) example of a double bond illustrated in ethylene (C_2H_4) molecule*

The geometric form with the three coplanar orbitals and the $2p_z$ orbital perpendicular to that plane is illustrated in Figure 1.12(b).

Double bond: in a carbon–carbon bond, as is found, for example, in the ethylene molecule C_2H_4 (Figure 1.12(c)), carbon is said to be double bonded, because the bond is formed of an s bond on the C–C axis, and a p bond.

1.4.6. *The sp hybridization of carbon*

In the case of *sp* hybridization, the $2s$ and $2p_x$ orbitals are reorganized to form two orbitals identical in shape, orientated along the same axis. The $2sp_y$ and $2sp_z$ orbitals, not involved in the hybridization, are perpendicular to the axis of these two aligned orbitals.

This results in a triple bond, comprising a sigma bond ($2sp$), and the two pi bonds ($2sp_y$ and $2sp_z$).

An example illustrating a triple bond between two carbon atoms is found in the acetylene molecule C_2H_2. An *s* bond is formed by the overlap of the bonds of the two *sp-sp* hybrid orbitals of each of the two carbon atoms and two p bonds by the overlap of each of the orbitals $2p_y$ (Figure 1.13(c)) and $2p_z$ of each of the two carbon atoms.

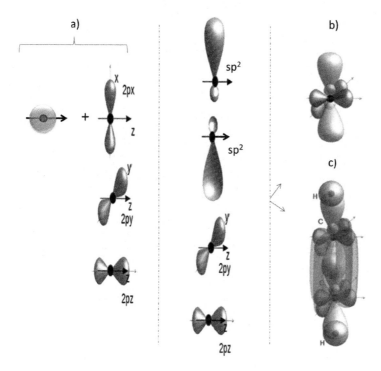

Figure 1.13. *Illustration of sp hybridization: a) reorganization of the atomic orbitals; b) orientation of the atomic orbitals around the carbon atom; c) example of a triple bond with the acetylene molecule C_2H_2, formed of a σ bond orientated along the C–C axis, and two π bonds orientated perpendicularly to that axis*

The different types of carbon bonds are summarized below.

Type of hybridization	Number of sp molecular orbitals	Number of 2p molecular orbitals	Number of s bonds	Number of p bonds	Type of bond
sp^3	4		1		Single
sp^2	3	1	1	1	Double
sp	2	2	1	2	Triple

Table 1.2. *Overview of the different kinds of carbon bonds*

In molecules with molecular orbitals s and p, it is the p molecular orbitals which condition the physio-chemical properties, and more specifically, the electrical and optical properties. To understand this, it is helpful to look at the level of energy in these different orbitals.

1.5. Energy diagram of different types of hybridization

By applying the principle of energy conservation, we can quite easily see that the energies of the hybrid orbitals are reduced in comparison to the energy of the $2p$ orbitals, but are still greater than the energy of the orbital $2s$. Indeed, if the total energy of the electrons in the layer $n = 2$, in the base state, is constant for the different hybridizations, it seems that:

– the total energy of the layer $n = 2$ is that of an electron in the orbital $2s$ and three electrons in the orbital $2p$, so: $E_{n=2} = E_{2s} + 3E_{2p}$;

– the energy associated with four electrons of the four sp^3 hybrid orbitals is the total energy. The value of this energy is:

$$E_n = 4E_{sp^3} \tag{1.2}$$

It can be deduced that the level of energy of an sp^3 electron is:

$$E_{sp^3} = \frac{\left(E_{2s} + 3E_{2p}\right)}{4} \tag{1.3}$$

– the energy associated with the three sp^2 electrons is that of a $2s$ electron and two $2p_x$ and $2p_y$ electrons:

$$3E_{sp^2} = E_{2s} + E_{2p} \qquad [1.4]$$

$$E_{sp^2} = \frac{\left(E_{2s} + 2E_{2P}\right)}{3} \qquad [1.5]$$

Hence, the level of the last quantum case is that of an electron in the remaining $2p$ orbital, so E_{2p}:

– the energy associated with two sp orbitals is that which is equivalent to a $2s$ orbital and a $2p$ orbital:

$$E_{sp} = \left(E_{2s} + E_{2p}\right)/2 \qquad [1.6]$$

The energy level of the two remaining quantum cases is E_{2p}.

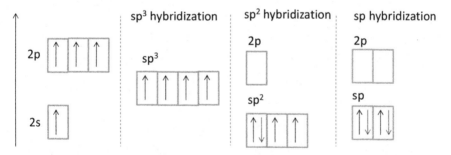

Figure 1.14. *Energy levels of the different electrons in layer n=2 for the different hybridizations. From left to right: carbon alone in the base state, sp^3 hybridization, sp^2 hybridization, and sp hybridization. The energy of the hybrid orbitals decreases in the order 2p, sp^3, sp^2, sp*

Because E_{2s} is lower than E_{2p}, the decreasing order of the energy levels is, respectively, the energy level E_{2p} of an electron in level $2p$, the energy level

E_{sp^3} of an electron in sp^3 hybridization, followed by the energy level E_{sp^2} of an electron involved in an sp^2 hybridization, and finally the energy level E_{sp} of an electron involved in an sp hybridization.

It is worth noting that the overlapping of the s molecular orbitals is more important than that of the p molecular orbitals. However, as illustrated by Figure 1.15, hybridizations involving s bonds have higher energy than p bonds. Hence, p bonds can be considered to be more "fragile" than s bonds, because it takes less energy to break them. It is helpful to note the link between these differences in s and p overlaps, on the one hand, and the solidity of the bonds and their energy level, on the other.

Note, also, in Figure 1.13, that the last occupied quantum case – what is known as the *Highest Occupied Molecular Orbital* (HOMO) – corresponds to a π bond. Similarly, the first empty quantum case, known as the *Lowest Unoccupied Molecular Orbital* (LUMO), is part of a π bond, or more specifically, a π* bond, as we shall see in the following few sections.

These molecular orbitals, HOMO and LUMO, play a crucially important role in the chemical properties and in the optical and electrical properties of organic semiconductors. To understand this, we need to look at the energy levels of the molecules containing more atoms and hybrid bonds. The energy levels π but also π* must be considered.

1.6. Conjugate molecules

The aim of this section is to show how the energy levels of the molecular orbitals evolve when the number of atoms increases or when the length of the molecule increases – i.e. when the number of hybridizations, and therefore of s or p bonds, increases.

In chemistry, a set of atoms bonded together by covalent bonds, at least one of which is a delocalized π bond, is called a conjugate system. We say that there is conjugation when a molecule contains an alternating series of single bonds (sp^3 hybridization and thus the presence of an s bond) and double bonds (sp^2 hybridization, and therefore the presence of both s and p bonds) – see Figure 1.15.

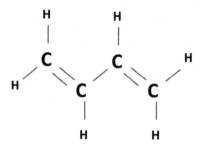

Figure 1.15. *Examples of a conjugate molecule: butadiene C_4H_6*

To illustrate the influence of the number of bonds on the energy levels, we shall consider the cases of ethylene and benzene, C_6H_6, with six bonds linking six carbon atoms.

1.6.1. *Ethylene*

In the case of ethylene, which served as an illustration for sp^2 hybridization of carbon, the two carbon atoms are therefore linked by an s bond doubled by a p bond which corresponds to the overlap of the $2p_z$ orbitals of each of the carbon atoms. The possibilities of combination of the wave functions ψ_1 and ψ_2 of each of the carbon atoms 1 and 2 are $(\psi_1 + \psi_2)$ and $(\psi1 - \psi_2)$, which correspond respectively to the two energy levels π and π^*, respectively at the bottom and top of Figure 1.16. In the base state, only the π level is occupied.

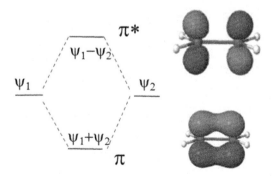

Figure 1.16. *Ethylene: on the left, energy levels of the π and π^* molecular orbitals, and on the right, representation of the signs of the wave functions for the $2p_z$ orbitals*

1.6.2. *Benzene*

Benzene (C_6H_6), shown in Figure 1.17, contains six carbon atoms, bonded together in a flat ring structure by an alternating series of single bonds (*sp* hybridization with one *s* bond and two *p* bonds) and double bonds (*sp*^{2} hybridization with one *s* bond and one *p* bond). The alternation of the three single and three double bonds leads to the combining of the atomic orbitals of the six carbon atoms A1, A2, A3, A4, A5 and A6 involved in the *p* bonds in six different ways, as illustrated in Figure 1.17. These six combinations correspond to the six possibilities for combining the alternation of signs of the wave functions Ψ_1, Ψ_2, Ψ_3, Ψ_4, Ψ_5 and Ψ_6:

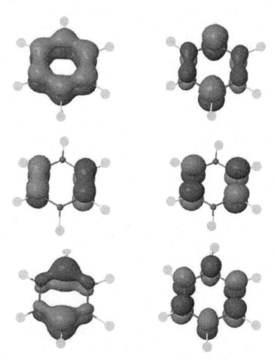

Figure 1.17. *The six possible combinations of the signs of the wave functions of the 2p_z orbitals of the benzene molecule*

There are six resulting molecular orbitals, giving six different energy levels p1, p2, p3, p4*, p5* and p6*, including three levels for bonding orbitals and three for antibonding orbitals.

1.7. Conjugate polymers

One of the typical examples of conjugate polymers is polyacetylene, represented in Figure 1.18. Polyacetylene is a long molecular chain, repeating the pattern C_2H_2 a great many times $[C_2H_2]_n$, and displaying an alternating pattern of single and double bonds (n may be in the tens – i.e. 10–60 double bonds). As the number of p bonds is very high, the number of energy levels π and π^* is also high, which may lead, no longer to a discrete number of levels, but instead to two quasi-continua, resulting in two energy bands. These two bands are, respectively, the equivalent of a valence band and a conduction band in inorganic semiconductors, as illustrated in Figure 1.18.

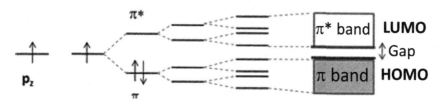

Figure 1.18. *Evolution of the number of energy levels as a function of the number of bonds. HOMO: Highest Occupied Molecular Orbitals. LUMO: Lowest Unoccupied Molecular Orbitals*

Organic semiconductors are classified on the basis of their size and their molar mass, into two main categories, which are identical in terms of physical behavior but differ in terms of the fabrication processes used. Small molecules, with a relatively low molar mass, can be deposited by vacuum evaporation. Polymers – molecules with a regular repeating pattern, forming long chains made up of elementary entities (monomers) – are employed by way of very inexpensive wet techniques.

1.8. Influence of conjugation length

Looking at Figure 1.19, we can see that the forbidden band between the HOMO and LUMO levels is reduced when the length of the conjugation chain increases. Hence, the length of the conjugation chain is one way of controlling the emission wavelength of a polymer.

To quantify the reduction in gap energy depending on the length of the molecular chain, a simple model of molecular orbitals of free electrons is sufficient: supposing we have a row of N atoms, spaced a distance d apart, the length of the chain is $L = (N-1)d$ ($L \sim Nd$ if N is very high). Quantum mechanics teaches us that a free particle (an electron) in a one-dimensional box (the molecular chain) experiences null potential inside the box and infinite potential outside it, so the wave function corresponding to the eigenvalues E_n is such that:

$$E_n = \frac{n^2 h^2}{8m(Nd)^2}$$
[1.7]

where $n = 1, 2, 3, \ldots$ represents the different rungs on the ladder, h is Planck's constant, m is the mass of an electron and n is an integer (quantum).

Supposing that the N electrons in the π orbitals occupy the levels of that "ladder", being arranged at two per level, and beginning by filling the lowest levels by virtue of the Pauli principle, the last electron will occupy the highest occupied molecular orbital (HOMO), and will have the energy:

$$E_{HOMO} = \left(\frac{N}{2}\right)^2 \frac{h^2}{8m(Nd)^2}$$
[1.8]

The level immediately above this would be the lowest unoccupied molecular orbital (LUMO), with energy:

$$E_{LUMO} = \left(\frac{N}{2}+1\right)^2 \frac{h^2}{8m(Nd)^2}$$
[1.9]

The prohibited band (band gap) between these two levels is simply the difference:

$$\Delta E = E_{LUMO} - E_{HOMO} = (N+1)\frac{h^2}{8m(Nd)^2}$$
[1.10]

As an initial approximation, this gives us:

$$\Delta E = \frac{h^2}{8md^2}\frac{1}{N}$$
[1.11]

This shows that when the number N of molecules increases to form a longer polymer chain, the band gap ΔE shrinks by $1/N$.

In the next section, we shall see that this very simple model does have certain limitations – particularly when the length of the molecular chain tends toward macroscopic dimensions.

As indicated above, the length of the conjugation chain is therefore a means of controlling a polymer's emission wavelength.

$$E_{gap} = h\frac{c}{\lambda} \qquad\qquad [1.12]$$

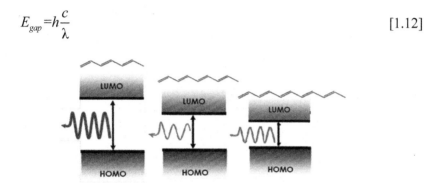

Figure 1.19. *Evolution of emission wavelength as a function of the length of the conjugate chain*

This property is highly interesting from the point of view of lasers, given the possibility of obtaining specific wavelengths as required, simply by varying the length of the organic chain. Consequently, organic materials are very advantageous for creating an organic laser. In the next section, we present the concept of organic molecules, and their optical and electronic properties.

1.9. Electronic properties of organic materials

As we have just seen, organic materials are made up, essentially, of molecules of carbon, hydrogen, nitrogen or sulfur. For example, those used in organic sources such as OLEDs (the subject of Chapter 2) are divided into two categories: polymers and materials with a low molecular mass (or "small molecules").

We saw above that the electron configuration of the base state of the carbon atom is divided between the 2s orbital and the three 2p orbitals (p_x, p_y, p_z). With organic materials, sp^2 hybridization is very significant. In this case, the 2s orbital and two of the three 2p orbitals are combined to form three sp^2 hybrid orbitals. The overlap of two sp^2 orbitals of the carbon atoms produces a strong π bond, whilst the linear combinations of atomic orbitals $2p_z$ gives rise to the π molecular orbitals (occupied bonding orbital which corresponds to the lowest energy band) and π* ones (unoccupied antibonding orbital which corresponds to the highest energy band) – see Figure 1.20.

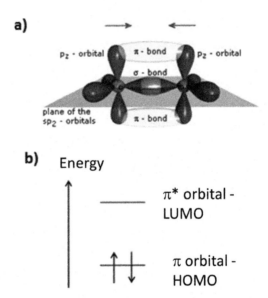

Figure 1.20. *(a) Distribution of σ and π bonds in a molecule of ethylene [www.orgworld.de]; (b) energy level of the molecular orbitals*

If the number of sp^2 hybrid carbon bonds increases, a system with an alternating series of single and double bonds forms. In a conjugate system (alternating single "C-C" bonds and double "C=C" bonds), the electrons can no longer be attributed to a specific C-C bond, so their wave function is delocalized across the whole of the conjugate molecule. This phenomenon gives rise to the intrinsic conductivity of light-emitting organic materials.

When the atomic orbitals combine, they create a set of molecular orbitals with distinct energies, thus forming energy bands. The energy bands which are most important in describing organic compounds are the HOMO and the LUMO (both defined above). The band between the HOMO and LUMO is the electron gap, which lends the compound its semiconductive nature.

Increasing length of the conjugate chain (carbon bonds) leads to increasing bandwidth, and thus to a gradual reduction in the HOMO-LUMO gap and the prohibited band of the corresponding material, which should therefore exhibit metallic-type conductivity. The injection of a charge into the conjugate chain occurs not by the addition of an electron to the LUMO, but by a local lattice deformation. However, the delocalization of the π electrons along the conjugate chain is generally not homogeneous (double bonds are generally shorter than single bonds). If a double bond is "weakened", the network undergoes a distortion, known as Peierls distortion [PEI 55]. This results in the deformation of the carbonate backbone, causing a localization of the π electrons and leading to the opening, at the Fermi level, of a prohibited band or band "gap".

Note that the width of the band gap is typically around 2–3 eV in organic materials.

The energy difference between the HOMO and LUMO can be used as a measure of the molecule's excitability: the smaller the energy difference, the more easily the molecule can be excited. We shall see later on that it is important to choose the structure wisely in order to obtain a light emission in the desired wavelength range.

1.10. Optical properties of organic semiconductors

1.10.1. *Fluorescence and phosphorescence*

When an organic molecule is in an excited state, following the absorption of a photon, it returns to the lowest energy level of the excited state by relaxation (a non-radiative process). The lifetime of these non-radiative relaxations is around 10^{-12} seconds. Note that the new vibrational level of the excited state is still higher than that of the base state.

The molecule then returns to the base state, either by internal or external conversion (collision with other molecules), or by the emission of light.

This emission, when it takes place, is called fluorescence.

In the case of phosphorescence, the rise to the excited state takes place with spin inversion and passage into a triplet state. The return to the base state occurs more slowly, with the emission of photons and a second spin inversion. The light emission therefore lasts much longer, even after the incident excitation is no longer present.

The molecule relaxes from its lowest vibrational energy level in the excited state to a vibrational energy level in the base state. The duration of this process is extremely short – around 10^{-9} s – and the emission wavelength depends solely on the relaxation into the base state from the excited state.

The loss of energy by vibrational relaxation means that in an organic molecule, the photoluminescence spectrum (and the electroluminescence spectrum) is shifted toward red (red-shifted) in relation to the absorption spectrum (Stokes shift), as shown in Figure 1.21.

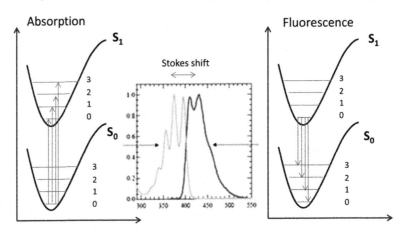

Figure 1.21. *The phenomenon of absorption and fluorescence in organic materials*

1.10.2. *Optical transitions in organic materials*

When an organic molecule is subjected to light excitation, it may absorb a photon. The molecule, which is initially in the electron energy base state, then moves into a higher energy electron state, known as the "excited" state. By the absorption of a photon, an electron–hole pair is generated. This process of absorption is efficient when the photon's energy is similar to the gap between the energy levels of the base state and the excited state. Such an excitation represents the passage of an electron from the base state (HOMO) to a higher-energy orbital (LUMO).

The quasi-particle formed by the electron (or the hole) and its (default) environment is called a negative polaron (or a positive polaron in the case of an electron hole). Coulombian interaction between an electron (or, specifically, a negative polaron) of the LUMO and a hole (a positive polaron) of the HOMO leads to the formation of what is known as an exciton.

The concept of an exciton plays a crucially important role in the physics of organic semiconductors and interactions in organic materials. There are two different types of excitons: singlet excitons and triplet excitons. A singlet exciton is characterized by the presence of an electron–hole pair with opposing spin, whereas the triplet exciton comprises an electron–hole pair with the same spin.

The physics of an excited molecule is described by the Born–Oppenheimer approximation [BAR 75]. This approximation explains that the motion of the electrons can be separated from that of the nuclei in view of the ratio between their masses. The Born–Oppenheimer approximation enables us to calculate the wave function $\Psi_{molecule}$ of a molecule in two consecutive steps: the electron component $\Psi_{electron}$ and the nuclear component $\Psi_{nuclear}$. In other words, the energy levels of a molecule can be determined by calculating the energies of the electrons for a given atomic position. By repeating this calculation for various atomic arrangements, the structure of the molecule at equilibrium can be determined, and the curve of the molecular energy potential as a function of the configuration can be plotted (Figure 1.22).

In this figure, an optical transition is represented by a vertical line, by virtue of the Franck–Condon principle [BAR 75]. This principle, which is the analog of the Born–Oppenheimer approximation for optical transitions, states that electron transitions are essentially instantaneous in comparison to the time characteristic of optical transitions. Therefore, if the molecule needs to move to a new vibrational level during the electron transition, that new vibrational level must be instantaneously compatible with the atomic position.

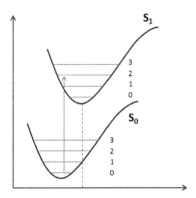

Figure 1.22. *Energy curve of the molecular potential for the base state S_0 and the excited state S_1*

Figure 1.22 shows the energy potential curves for states S_0 and S_1. When a photon is absorbed, an electron transition of the vibrational level V_0 from the base state to the excited state occurs, as indicated by a vertical line linking the two states of excitation. The excitation causes a modification of the molecule's geometric form. This geometric change is indicated by a change in the atomic coordinates between states S_0 and S_1.

According to the Franck–Condon principle [CLA 07], which indicates that the electron transition is sufficiently quick to take place with no alteration of the nuclear coordinate of the configuration associated with the position of the nuclei/atoms, the probability that the molecule will be found at a particular vibrational level is proportional to the square of the overlap between the vibrational wave functions of the initial and final states. For this particular example, illustrated in Figure 1.24, this means that the vibrational transition from V_2 to V_0 in the state S_1 is the greatest. After the electron transition, the molecule relaxes rapidly to its lowest vibrational level, and hence it can become de-excited and return to the state S_0 by the emission of a photon. The Franck–Condon principle also applies to absorption and fluorescence. Consequently, the emission spectrum is red-shifted in relation to the absorption spectrum. This red shift is known as the Stokes shift, and it varies from 0.1 eV to a few eV for organic materials [HAY 95].

As we have seen, it is possible to excite organic molecules by the absorption of a photon at a definite energy.

The Jablonski energy diagram [VAL 12] visually displays the different energy levels and energy transitions resulting from the absorption of a photon by an organic molecule (see Figure 1.23). In this diagram, the states are represented vertically in accordance with their energy level, and grouped horizontally on the basis of their multiplicity. A distinction must be drawn between singlet and triplet excitons. A singlet exciton is characterized by the presence of an electron–hole pair with opposing spin, whilst the triplet exciton comprises an electron–hole pair with the same spin. Because of the repulsive nature of the spin–spin interaction between electrons with the same spin, the triplet state has lower energy than the corresponding singlet state.

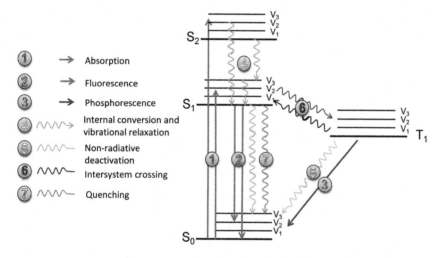

Figure 1.23. *Jablonski energy diagram. For a color version of this figure, see www.iste.co.uk/boudrioua/lasers.zip*

In this diagram, the singlet and triplet states are represented, respectively, by S_i and T_i, where i denotes the energy level. S_0, represented by a thick line, is the base state, which is a singlet state for all organic compounds. The thin lines represent the vibrational states. The radiative transitions are represented by a straight arrow, and non-radiative transitions by a wavy arrow.

Molecules at rest in the vibrational level V_0 of the electron base state S_0 are raised to an excited state S_1 by the absorption of light radiation. Such a state of excitation can release its excess energy by different de-excitation mechanisms. The first is the emission of a photon. The de-excitation of the molecule can occur by radiative transition to the different resonance levels of the base state having the same multiplicity (excited singlet S_1 – base singlet S_0). The emission spectrum obtained is called fluorescence. Fluorescence results in radiation bands, because the excited molecules can return to whatever vibrational level in the electron base state S_0. The lifetime of the singlet state S_1 is very short; of the order of 10^{-9} to 10^{-7} s. In the other case, the excited state can evolve, becoming a triplet state in a non-radiative transition called Inter-System Crossing.

The light emission which occurs between two states of different multiplicity (triplet T – base singlet S_0) is called phosphorescence. This process is much longer than fluorescence, of the order of 10^{-3} to 10^2 s. In addition, the transition from the triplet state to the base state is generally a non-radiative relaxation because of its long lifetime and the possible interactions with the other molecules in contact.

There are other levels of singlet states reached by the molecule when it is excited. When the molecule passes from a higher excited state to a lower excited state, the energy is transformed into heat. This process is called internal conversion (IC), and occurs in a very short period of time, of the order of 10^{-14} to 10^{-11} s between levels S_2 and S_1, given the slight difference in their energy. In addition, it is possible to have vibrational relaxations in an excited state to return to the lowest vibrational level (V_0). This process is very quick ($<10^{-12}$ s) and non-radiative.

The last case is also a non-radiative process, called "quenching". A de-excitation may occur following the collision between an excited molecule and another molecule. This is one of the main causes of decrease of fluorescence in organic materials. We shall come back to it in the section on losses.

1.10.3. *Energy transfer phenomena*

In addition to photoluminescence in organic materials, organic molecules exhibit the peculiarity of transferring energy, little by little, over distances of a few nanometers. This exchange mechanism is particularly interesting at high levels of excitation. An excited molecule (considered to be a donor D*) can pass its excitation energy to a neighboring molecule (the acceptor A) in a non-radiative fashion, by coupling between the electron orbitals of the two molecules. This phenomenon is known as resonance energy transfer, and is described by the following mechanism:

$$D^* + A \rightarrow D + A^*$$
[1.13]

This phenomenon can be used to improve light emission performance, making host–guest systems. Thus, by optical pumping in a doped organic layer, excitation can be transferred from a so-called "host" molecule to a "guest" molecule (the doping agent).

This expresses the quenching of the donor's fluorescence and an increase in the acceptor's fluorescent emission. In organic compounds, this energy transfer may take place by different Coulombian mechanisms of interaction (Förster long-range resonant energy transfer), or by overlap of the orbitals of D and A (short-range Dexter energy transfer) [MOL 11].

1.10.4. *Förster mechanism*

The Förster mechanism, described in 1948 [FÖR 48], is a non-radiative resonant dipole–dipole transition between singlet excited states of two molecules. The mechanism known as FRET (Fluorescence Resonance Energy Transfer) is illustrated in Figure 1.24. An excited "donor" molecule (D*) transfers its energy to a neighboring "acceptor" molecule (A). Förster transfer is described by Fermi's golden rule. The transfer rate Γ_{mn} from a donor state m to an acceptor state n is given by:

$$\Gamma_{m,n} = 4\pi^2 \left| V_{m,n} \right|^2 \delta(E_n - E_m)$$
[1.14]

where $V_{m,n}$ is the transfer matrix between the two states m and n, with energy levels E_m and E_n respectively. In view of this equation, we can see that the energy of two states must be balanced in order for the transfer to take place, as indicated by the Dirac δ function.

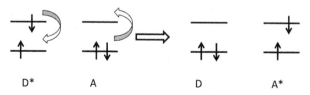

Figure 1.24. *Diagram of the principle behind the Förster mechanism*

Consequently, a precondition for Förster transfer is a spectral overlap of the emission spectrum of the donor and the absorption spectrum of the acceptor (see Figure 1.25). When we evaluate the elements of the interaction matrix and integrate over the donor and acceptor spectra, the Förster equation becomes [FÖR 48]:

$$\Gamma_{DA} = \frac{1}{\tau_D} (\frac{R_0}{R})^6 \qquad\qquad [1.15]$$

where R is the distance between the donor and the acceptor, τ_D is the lifetime of the donor's excited state and R_0 is the Förster radius, which is given by [FÖR 48]:

$$R_0 = \left(8.8 \times 10^{17} \frac{k^2}{n^4} \int g_A(E) g_D(E) dE \right)^{1/6} \qquad [1.16]$$

In this equation, κ is a coefficient which depends on the dipoles' orientation; n is the refractive index; and g_A and g_D are the absorption spectrum of the acceptor and emission spectrum of the donor, respectively.

For organic materials, Förster transfer is a long-range mechanism. It decreases as a function of R^{-6} and occurs over distances ranging from 2nm to 10nm.

The mechanism of Förster transfer is important, particularly for energy transfer in guest–host systems. Several bi-molecular annihilation processes, occurring in organic materials at high excitation and whose impact is particularly decisive for organic laser diodes, are based on Förster transfer.

Förster transfer requires the transitions of the donor and acceptor to be authorized; in other words, energy transfer between triplet states is prohibited because of the condition of spin conservation of each type of molecules in dipole–dipole interactions. Therefore, only transitions employing singlet excitons are authorized.

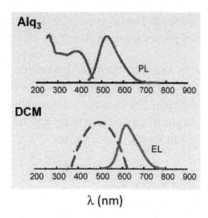

Figure 1.25. *Representation of the absorption- and emission spectra for a thin layer of Alq3 and DCM*

1.10.5. *Dexter mechanism*

Dexter transfer is a mechanism whereby the exciton diffuses from a donor excited state to an acceptor excited state [DEX 53]. Unlike Förster transfer, where energy is transferred by resonance, Dexter transfer is a transfer of charges between the donor and acceptor. It results from the transfer of an excited electron from the LUMO in the donor to the acceptor, and *vice versa*, the transfer of a non-excited electron from the HOMO in the acceptor to the donor by the spatial overlapping of the orbitals (Figure 1.26). The same mechanism can be applied to the transfer of triplet excitons.

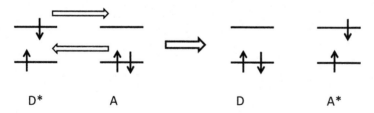

Figure 1.26. *Diagram of the principle of the Dexter mechanism*

The rate of Dexter transfer is proportional to e^{-2R}, where R is the distance between the donor and acceptor [DEX 53]. Because the Dexter transfer rate is an exponential function of the distance between the donor and acceptor, the Dexter mechanism depends heavily on the transfer distance. This transfer mechanism takes place at short range, of the order of 0.5 to 2 nm.

To summarize, energy can be transferred between singlet state excitons by energy transfer by the Förster mechanism, whilst the Dexter mechanism describes the transfer of charge. Both transfer mechanisms play an important role in optoelectronic organic devices, where the bi-molecular interaction losses are significant at high levels of excitation.

1.11. Losses in organic materials

1.11.1. *Bi-molecular interaction losses*

Bi-molecular reactions are connected to interactions in the organic material. The direct consequence of this process is the decrease of the fluorescence, which is a limitation for light emission. Losses by bi-molecular interaction occurs between charge carriers and excitons. Thus, these losses are particularly significant at high levels of excitation, where the densities of the charged particles are high. These losses are connected to processes which have a negative impact on the density of the singlet excitons, which influences the threshold of organic laser diodes. In this section, the physical context and fundamental mechanism of losses by bi-molecular interaction are presented.

1.11.2. *Losses by polaron absorption*

As indicated in the previous sections, electrical excitation involves the injection of a charge. The charges injected and their polarization field are called polarons. They may cause additional emission losses by absorption, known as polaron absorption losses [KOZ 00], and can lead to the annihilation of excited states by interaction with polaron states, as shown by the equations below:

$$S_1 + n \xrightarrow{\ K_{SPA}\ } S_0 + n \qquad\qquad [1.17]$$

$$T_1 + n \xrightarrow{\ K_{SPA}\ } S_0 + n \qquad\qquad [1.18]$$

where S_1 and T_1 are, respectively, the first singlet and triplet excited states, and n represents the charge.

1.11.3. Singlet–singlet (S–S) losses

Losses through singlet–singlet interaction describe the energy transfer from one singlet exciton to another by collision (known as singlet–singlet annihilation, or SSA) [NAK 05]. The reaction equation for SSA is given by:

$$S_1 + S_1 \xrightarrow{\ k_{SSA}\ } S_1 + S_0 \qquad\qquad [1.19]$$

where the reaction rate coefficient is written as k_{SSA}. The first excited singlet state and base state are written as S_1 and S_0, respectively. The reaction rate k_{SSA} can be obtained by transient photoluminescence measurements [SOK 96].

1.11.4. Triplet–triplet (T–T) annihilation

The interaction of two triplet excitons (T–T) also leads to non-radiative energy transfer between the first and second exciton.

$$T_1 + T_1 \xrightarrow{\ k_{TTA}\ } T_2 + S_0 \qquad\qquad [1.20]$$

where k_{TTA} is the rate of triplet–triplet interaction. T_1 and T_2 are, respectively, the first and second excited level of the triplet state.

The excited state T_2 relaxes from the first singlet or triplet state. This results in the formation of a singlet exciton or a triplet exciton. The spin statistics determine the probability of formation of the singlet and triplet exciton [REU 05]. The ratio of the singlet exciton to the number of triplet excitons is described by the parameter ξ. Three times more triplet excitons than singlet excitons are generated, giving a probability that the exciton will be a singlet of $\xi = 0.25$ [BAL 99]. The final reaction equations, for the case of triplet–triplet annihilation, are given by:

$$T_1 + T_1 \xrightarrow{\ (1-\xi)k_{TTA}\ } T_1 + S_0 \qquad\qquad [1.21]$$

$$T_1 + T_1 \xrightarrow{\ \xi k_{TTA}\ } S_1 + S_0 \qquad\qquad [1.22]$$

The quantum yield is limited mainly by the fact that only the singlet states are radiative in fluorescent materials.

1.11.5. *Singlet–triplet (S–T)*

Losses are also caused by the interaction of singlet and triplet excitons (singlet–triplet annihilation, or STA). Given the spin nature of phosphorescence, the radiative lifetime of triplet excitons is greater than that of singlet excitons. Consequently, the number of excitons in the triplet state accumulates, leading to new losses. In this process, the energy of a singlet exciton is transferred to a triplet exciton, which is excited to a higher triplet state. Then, the excited triplet exciton quickly relaxes by internal conversion to a lower excited triplet state. The coefficient is called k_{STA}. The equation for this process is:

$$S_1 + T_1 \rightarrow T_2 + S_0 \xrightarrow{k_{STA}} T_1 + S_0 \qquad [1.23]$$

In this process, a singlet exciton is eliminated, transferring its energy to a triplet exciton. The reverse process – i.e. the transfer of energy from a triplet exciton to a singlet exciton – is prevented by the spin. However, in a phosphorescent emitter with strong spin–orbit coupling, this process can take place:

$$T_1 + S_1 \xrightarrow{k_{STA}} S_0 + S_{n\geq1} \rightarrow S_0 + S_1 + heat \qquad [1.24]$$

1.11.6. *Losses by inter-system crossing*

Sometimes, we see non-radiative transitions between different spin states. This phenomenon is known as inter-system crossing (ISC). In organic materials, the singlet exciton can be efficiently transformed into a higher vibrational state than the triplet state. Next, the singlet exciton relaxes into the lowest triplet state. The process describing ISC is discussed in detail in the section on optical transitions (section 1.11.2), where the constant of rate of inter-system crossing is written as k_{isc}.

$$S_1 \xrightarrow{k_{ISC}} T_1 \qquad [1.25]$$

A typical case of reduction of the quantum efficiency is observed, when an organic material is optically pumped to produce a stimulated emission. When the density of excited molecules increases, it is common to observe a reduction in quantum efficiency of fluorescence. Thus, an increase in the likelihood of interaction between molecules leads to a reduction in the number of singlet excitons.

1.11.7. *Polaron absorption losses*

Electrical excitation involves the injection of a charge. The charges injected and their polarization field are called polarons. Polarons can cause additional absorption losses called polaron absorption losses [KOZ 00], and may lead to the annihilation of the excited states by interaction with the polaron states, as shown by the below equations:

$$T_1 + n \xrightarrow{k_{STA}} S_0 + n^* \qquad\qquad [1.26]$$

$$S_1 + n \xrightarrow{k_{STA}} S_0 + n^* \qquad\qquad [1.27]$$

1.12. Notions of photometry

The aim of visual photometry is to quantify the stimulation of an observer's eye by a light source. Its field of application is limited to the visible portion of the electromagnetic spectrum. Because the stimulation of the human eye varies from one individual to another, and depends on multiple conditions of observation, the CIE (*Commission Internationale de l'Éclairage* – International Committee on Illumination) defined a system of units based on the "CIE eye of a standard observer" (CIE 64) [PAL 68].

Experimentally, we see that the eye is sensitive to light radiation at wavelengths between 400 and 800 nm, but it does not have the same spectral sensitivity. Indeed, the sensitivity of the human eye is greater at the center of the visible spectrum and decreases on both sides. The standard spectral sensitivity curve $V(\lambda)$ of daytime vision (termed photopic vision) is maximal at 555 nm, as shown in Figure 1.27.

The curves in this figure are important in determining the visual photometric values, which are the subject of the following paragraphs.

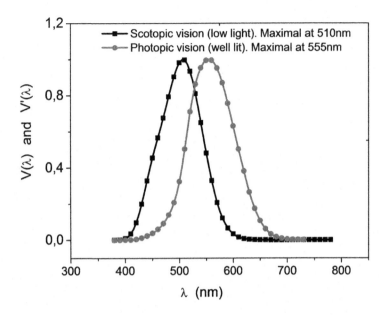

Figure 1.27. *Standard CIE curve of spectral sensitivity of the human eye V(λ)*

1.12.1. *Light flow*

The light flow Φ_v is the basic photometric value which characterizes the total amount of electromagnetic radiation emitted by a source, spectrally weighted by the spectral luminous efficiency function of the human eye $V(\lambda)$. The light flow unit is the Lumen (Lm). The source is, in principle, considered to be monochromatic, with the wavelength λ, which corresponds to the most intense emission (generally situated near to the center of the spectrum).

In fact, the majority of organic materials exhibit a broad emission spectrum and the spectral distribution of the energy flux Φ_e is defined on the basis of the spectral energy flux $\Phi'_e(\lambda)$ such that:

$$\lim_{\delta\lambda \to 0}\left(\frac{\delta\Phi_e}{\delta\lambda}\right) = \Phi'_e(\lambda) \qquad [1.28]$$

where $\delta\Phi_e$ represents the fraction of Φ_e situated in the spectral band of breadth $\delta\lambda$ taken around the wavelength λ.

The total flux corresponding to the spectral band $[\lambda_a, \lambda_b]$ is then given by:

$$\Phi_e(\lambda_a, \lambda_b) = \int_{\lambda_a}^{\lambda_b} \Phi_e'(\lambda).d\lambda \qquad [1.29]$$

Conversely, to calculate $\Phi_e'(\lambda)$ based on the total flux Φ_e, we use the normalized electroluminescence spectrum $S(\lambda)$ of the OLED.

In fact, $\Sigma = \int_0^\infty S(\lambda).d\lambda$ represents the area of the normalized spectrum and the total flux can be given by:

$$\Phi_e = a.\Sigma$$

where a = constant.

Thus, we have $\Phi_e'(\lambda) = a.s(\lambda)$, and finally:

$$a = \frac{\Phi_e}{\Sigma} = \frac{\Phi_e'(\lambda)}{S(\lambda)} \qquad [1.30]$$

On the basis of $S(\lambda)$ and Φ_e, we can then write:

$$\Phi_e'(\lambda) = \frac{\Phi_e.S(\lambda)}{\Sigma} = \frac{\Phi_e.S(\lambda)}{\int_0^\infty S(\lambda).d\lambda} \qquad [1.31]$$

For a monochromatic source (emission at λ_d), the relation between the energy flux Φ_e and the light flux Φ_v is:

$$\phi_v(\lambda_d) = K_m V(\lambda_d)\phi_e(\lambda_d) \qquad [1.32]$$

where K_m = 683 lm/W, and where $V(\lambda)$ represents the normalized photonic response of the eye.

For a polychromatic source:

$$\Phi_v = \Phi_v(\lambda_s, \lambda_d) = \int_{\lambda_d}^{\lambda s} \Phi'_v(\lambda_d)d\lambda \qquad [1.33]$$

$\Phi'_V(\lambda)$ is the spectral luminous flux such that: $\Phi'_V(\lambda) = K_m.V(\lambda)\Phi'_e(\lambda)$, and finally:

$$\Phi_V = K_m.\int V(\lambda).\Phi'_e(\lambda).d\lambda \qquad [1.34]$$

1.12.2. *Luminous intensity*

The luminous intensity (for a point source in a given direction) is the luminous flux per unit solid angle in the direction in question. Its unit is the candela (cd). To calculate the luminous intensity, we very often consider an OLED to be a Lambertian source [RIE 15]. The intensity of the flux per unit solid angle (the flux per stereo-radian) defines the luminous intensity "I_v". For a Lambertian source, the total flux emitted is linked to the intensity by:

$$\Phi_v = 2\pi \int_0^{\pi/2} I_v(\theta)\sin(\theta)d\theta \qquad [1.35]$$

where $I_v(\theta) = I_{v,0}\cos(\theta)$ $\qquad [1.36]$

$I_{v,0}$: the maximal intensity perpendicular to the emission surface

The solution to the previous equation shows that the luminous flux of a Lambertian source is $\pi I_{v,0}$.

For that source, the luminous intensity is directly proportional to the cosine of the angle between the observer and the normal to the emission surface (Figure 1.28).

The previous approximation considering the OLED to be a Lambertian source must be taken with great caution. Although it is widely used, this approximation is, unfortunately, often far removed from reality. The OLED is a very fine emission source, and it is composed of different thin layers, both organic and inorganic (the electrodes). Hence, it is the site of significant interference effects which can greatly influence the emission spectrum. This is especially true in the case of OLEDs possessing the geometry of a microcavity [COE 13], which causes the emission diagram to differ greatly from the Lambertian profile (Figure 1.28).

In spite of these considerations, we continue to use the approximation of a Lambertian source to calculate luminous intensity and luminance.

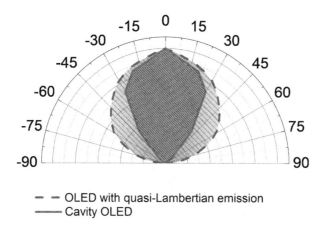

— — OLED with quasi-Lambertian emission
—— Cavity OLED

Figure 1.28. *Dashed line: emission diagram of a Lambertian source. Solid line: emission diagram of an OLED, with a very prevalent microcavity effect*

1.12.3. *Luminance*

Luminance (a point on a surface in a given direction) is interpreted as the luminous intensity perceived in a given direction and emitted by a surface element of the source. It is measured in cd/m². Luminance can be calculated on the basis of the luminous intensity for $\theta = 0°$, and it is given by:

$$L_v = \frac{I_v(\theta)(\text{for } \theta = 0)}{S} = \frac{I_{v,0}}{S} \qquad [1.37]$$

where S is the active surface area of the OLED.

The luminance can also be easily deduced from the luminous flux:

$$Lv = \frac{\phi_v}{\Omega.S_a}$$ [1.38]

where Ω is the solid angle of emission.

1.12.4. *Illumination*

Unlike luminance, illumination can be defined as the luminous flux per unit surface of a receiver or sensor. Its unit is the lux (Lx) or (lm/m²). Illumination can be calculated using the following relation:

$$E = \frac{d\Phi}{dS}$$ [1.39]

Table 1.3 gives an overview of the photometric and visual values and their energetic equivalents.

Values	Definition	Energy units	Visual units	Photonics
Energy	nhv	Joules	Lumen.s	$Q = hv$
Flux	$\varphi = \frac{dQ}{dt}$	W	Lumen	Number of photons/ second
Intensity	$I = \frac{d\varphi}{d\Omega}$	W/sr	Lumen/sr (candela)	Number of photons per second per steradian
Luminance	$L = \frac{dI}{dS}$	W/m².sr	Lumen/m².sr = cd/m²	Number of photons per unit surface and per steradian
Illumination	$E = \frac{d\varphi}{dS}$	W/m²	Lumen/m² (lux:[lx])	Number of photons per unit surface and per second

Table 1.3. *Summary of a few photometric and radiometric values*

1.12.5. *Yields*

There are multiple definitions to calculate the efficiency of an organic source. The choice between the different definitions depends on the type and the intended domain of application for the light source.

1.12.5.1. *Quantum yield*

In the case of an electrical excitation, the quantum yield characterizes the number of photons N_{phext} emitted by the source in the outer half-space, in relation to the number of electrons injected N_{el}. It is obviously a radiometric efficiency which takes account of the radiometric values only. It is given by:

$$\eta_{ext} = \frac{N_{phext}}{N_{el}}$$

[1.40]

The number of electrons injected into the organic hetero-structure can be calculated on the basis of the electrical power absorbed:

$$N_{el} = \frac{P}{E_e}$$

[1.41]

E_e is the energy of an electron.

The number of electrons injected, therefore, is:

$$N_{el} = \frac{P}{E_e} = \frac{UI}{qU} = \frac{I}{q}$$

[1.42]

The number of photons emitted is calculated on the basis of the energy flow. In the case of a monochromatic source:

$$N_{phext} = \frac{\varphi_E}{E_{ph}} = \frac{\pi \lambda L_E S_D}{hc}$$

[1.43]

The quantum efficiency is therefore:

$$\eta_{ext} = \frac{\pi \lambda L_E S_D q}{hcI}$$

[1.44]

In the case of a polychromatic source, we can show that the quantum efficiency is given by:

$$\eta_{ext} = \frac{\pi S_D q}{hcI} \int \lambda L'_E(\lambda) d\lambda \qquad [1.45]$$

1.12.5.2. Internal quantum yield η_{int}

The internal quantum yield characterizes the total number of photons emitted in relation to the number of electrons injected. This yield characterizes the efficiency of the processes involved. It is hugely important in optimizing performances, but is difficult to calculate and measure. Indeed, we can view it as being proportional to the external quantum yield, but the proportionality factor is complex to determine. In the case of a source such as an OLED (which we shall see in the next chapter), we must take account of:

– the losses caused by the reflection of the emitted light between the anode and the cathode;

– the refraction indices of the different organic layers (emissive, transport of electrons or holes, etc.);

– the absorption of the ITO anode (between 15% and 20% depending on the range of wavelength).

1.12.5.3. Light yield

Although this efficiency is well known, it is often more common, in the case of OLEDs, to use photometric efficiencies which take into consideration the human perception of the light source.

The yield of light power expressed in lm/W is the most widely used photometric yield. It is given by the ratio between the luminous flux emitted by a light source and the power absorbed by that source.

If we let P represent the power received by the source (usually in the form of electricity for LEDs) and Φ is the luminous flux emitted, then the light yield η_L, by definition, has the value:

$$\eta_v = \frac{\Phi_v}{P} = \frac{\pi . L_v . S_D}{U.I} \qquad [1.46]$$

where:

  – P: the electrical power dissipated in the OLED (W);

  – U: the applied voltage across the OLED (V);

  – I: the electrical intensity injected into the OLED (A);

  – S_D: the emitting surface area of the OLED (m^2);

  – L: the luminance (Cd/m^2).

It is also very common to use a different photometric efficiency: luminous efficiency. It is defined as the ratio between the luminance of the OLED and its current density, and so is expressed in cd/A.

1.13. Concepts of colorimetry

Another essential tool in characterizing the emission of an OLED is to determine the chromatic coordinates of the light emitted. The human eye is capable of discerning over 350,000 different colors. To simply and effectively characterize colors, it is therefore necessary to classify them in a system that is immune to the imperfections of the human eye and the defects in its function. It is necessary to use a physical approach such as the RGB system (i.e. Red, Green and Blue).

Any color can be represented by a vector of trichromatic components X, Y, Z such that [SCH 07]:

$$X = K_m \int L_e \bar{x}(\lambda)d\lambda$$
$$Y = K_m \int L_e \bar{y}(\lambda)d\lambda \qquad\qquad [1.47]$$
$$Z = K_m \int L_e \bar{z}(\lambda)d\lambda$$

$\bar{x}(\lambda)$, $\bar{y}(\lambda)$, $\bar{z}(\lambda)$ are called colorimetric functions.

From the trichromatic component vector, we deduce the normalized trichromatic coordinates x, y, z such that:

$$x = \frac{X}{X+Y+Z}, \ y = \frac{Y}{X+Y+Z} \ \text{and} \ z = \frac{Z}{X+Y+Z} \qquad [1.48]$$

As $x + y + z = 1$, we represent the colors in a two-dimensional framework (xy) with orthogonal axes. This framework is called the chromaticity diagram (see Figure 1.29).

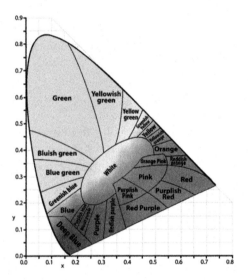

Figure 1.29. *Chromaticity diagram, from CIE 1931. For a color version of this figure, see www.iste.co.uk/boudrioua/lasers.zip*

This system exhibits a number of advantages:

– the trichromatic components are always positive;

– the white color has the trichromatic coordinates (1/3, 1/3, 1/3);

– we only need to know two out of the three coordinates to determine the position of the chromatic point.

1.14. Conclusion

This first chapter recaps the electronic properties of organic materials, and more specifically, organic semiconductors, which condition the mechanisms of charge injection and mobility, and the phenomenon of conduction. Those same electronic properties at the molecular scale enable us to describe the mechanisms of transfer and energy losses, radiative or non-radiative recombinations in organic materials. These phenomena, based on the excited states lifetime (or annihilation rates) are significant, and influence the dynamic of the laser effect sought in organic materials.

2

Organic Light-emitting Diodes

The earliest work on electroluminescence of organic materials is often attributed to Tang and Van Slyke, from the Kodak laboratories in Rochester, New York [TAN 87]. However, in 1953, the Frenchman André Bernanose from the Faculty of Pharmacy in Nancy presented work on the electroluminescence of acridine [KAL 96], and Pope, in 1963, reported electroluminescence in a crystal of anthracene [SHO 65]. However, this pioneering work produced very little in the way of results, because the electrical devices used were capacitors, rather than diodes. Indeed, in a structure made up of an organic monolayer sandwiched between two electrodes, the difference in mobility between the holes and electrons gives rise to recombination, not in the heart of the material electroluminescent, but mainly in a region near to the electrode injecting the minority charge carriers, i.e. near to the electrode. The efficiency of electric current-to-light conversion, therefore, is very poor.

The first organic light-emitting diodes (OLEDs) succeeded in overcoming the poor efficiency observed in earlier attempts thanks to the use of OLED heterostructures composed of different thin organic films. The main advantage of an OLED heterostructure is the possibility of bringing about a recombination of the charges in an emissive layer having high fluorescent efficiency. Charges of opposing signs, coming from the two electrodes, move through the injection layers and charge transport layers, meet and then recombine in an emission zone ideally situated in the very center of the OLED heterostructure.

The sections below lay down the foundations necessary to understand how OLEDs work, and the various physical processes involved.

2.1. Operation of an OLED

In order for the charges to recombine far from the electrodes, and for the heterostructure to be reasonably conductive, the materials from which the heterostructure is made must satisfy two criteria: a high charge mobility in the different organic layers, and an alignment of the HOMO and LUMO energy levels, which favors the channeling of charges between the different layers. In other words, the heterostructure must contain layers whose role is to facilitate the injection of charges from the electrodes, and their subsequent transport to the recombination or emission zone. In addition, to force recombination far from the electrodes, a blocking layer may be used.

Figure 2.1 shows an OLED heterostructure with the various layers mentioned above.

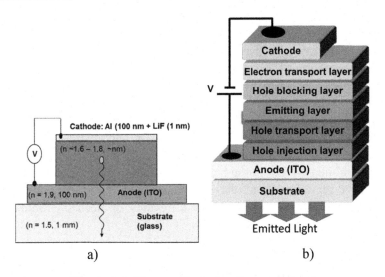

Figure 2.1. *Diagram illustrating the principle of a) a simple OLED and b) an OLED heterostructure*

An OLED works, essentially, in the same way as does an inorganic light-emitting diode (LED). Electrons are injected by a low work-function cathode, whilst holes are injected by a high work-function anode. These charge carriers then come together in the emission layer, thanks to the electrical field applied between the two electrodes, and their recombination gives rise to the phenomenon of electroluminescence. The color of the emitted light can be adjusted by choosing the appropriate polymer or small molecule.

The phenomenon of electroluminescence in an OLED can be broken down into a number of successive steps, as represented in Figure 2.2:

– injection of holes into the HOMO band and electrons into the LUMO band;

– movement of charge carriers within the material;

– electron–hole recombination;

– light emission.

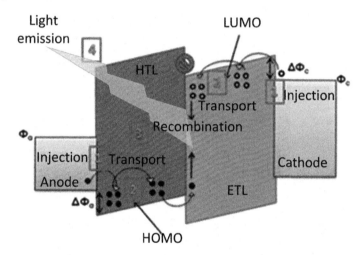

Figure 2.2. *Principle of light emission by an OLED device. The notations are as follows: Φ a for the anode work function; Φc for that of the cathode; ΔΦa and ΔΦc for the energy barriers that must be crossed by the holes and electrons respectively; Ea for the external electrical field; and hu for the light energy emitted. HTL: Hole Transport Layer and ETL: Electron Transport Layer*

The different factors limiting the efficiency of an OLED are illustrated in Figure 2.3:

– γ is the double charge injection factor;

– η_r is the rate of production of radiative-state excitons. It is connected to the statistic of spin for the production, in electroluminescence, of singlet states ($\approx$ 25% of excitons formed);

– η_f is the quantum yield of fluorescence. It is always less than 1 (0.9 in highly fluorescent material), because of the non-radiative recombinations that may occur (Jablonski diagram);

– η_{opt} is the optical coefficient introduced to take account of the refraction occurring at the diode/air interface. Only internal emissions hitting the interface at an angle less than a given limiting angle are able to be outcoupled from the OLED, and we obtain $\eta_{opt} = 1/2n^2$ for flat structures (with n being the refractive index of the emissive organic material).

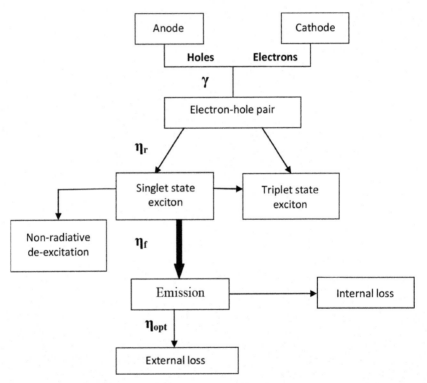

Figure 2.3. *Diagrammatic representation of the electroluminescence process*

The external quantum yield, which is the light yield, is the product of the different yields mentioned above – specifically:

$$\mu_{ext} = \frac{\text{Number of photons emitted outside the heterostructure}}{\text{Number of charges injected}} = \eta\mu_r \ \mu_f \ \mu_{opt} \quad [2.1]$$

Typically, that yield is estimated at between 3 and 4%.

The design of a high-performing light-emitting heterostructure therefore consists of, firstly, finding appropriate materials for the injection and transport of the charges (taking account of the HOMO, LUMO and the mobilities of the electrons or holes) and, secondly, optimizing the thicknesses of those layers as a function of the mobility of the materials.

2.2. Injection of charge carriers

When an OLED is working, the charge carriers are injected by the electrodes, and are transported through the various organic layers toward the opposite electrode. Along their path, the charges cross numerous interfaces. At each interface, the charge carriers may encounter an energy barrier which they need to surmount. Each energy barrier increases the OLED's working voltage, and also results in the manifestation of an area of charge accumulation, increasing the probability of non-radiative recombination [HUG 05]. Hence, the energy barriers at the interfaces cause higher working voltages and poorer efficiency.

With molecular engineering, which has meant we have a wide range of organic molecules from which to choose, it is possible to control, or even remove, the energy barriers between the different organic layers, by carefully choosing the organic semiconductors used in the OLEDs [KAZ 99, KOC 07, KAM 06, XU 06, ISH 99]. However, experience has shown that it is much harder to adjustment the interfaces between the electrodes and the organic layers. The electrodes are generally made of a transparent conductive oxide (TCO), such as indium tin oxide (ITO), aluminum-doped zinc oxide (AZO), fluorine-doped tin (di)oxide (FTO), or a metal such as Al or Ag. The energy barriers at the interfaces between organic materials and electrodes tend to be high.

In order to attenuate the problems due to energy barriers to the injection of charges at the electrodes, it is often necessary to use charge injection layers, interposed between the electrodes and organic layers. These injection layers may be thin "buffer" layers (around 0.1–10nm), made of oxides and/or alkyl halides (of LiF [CHA 13], CsF [PIR 00, JAB 00, CHA 04] or Li [PAR 01]), or thicker layers made of organic materials to inject/transport charges. They are used to control the alignment of energy levels, and are currently the subject of a great deal of research and debate.

Although the results are convincing in the case of the use of thin buffer layers, the mechanism to lower the energy barrier is, indeed, still a matter for some debate. For example, the influence of alkyl halides is complex, and the reaction mechanisms appear to be different depending on the type of material surrounding the buffer layer [JAB 98]. "Conventional", thicker injection layers are mainly used to reduce energy barriers by aligning the energy levels in the charge-transporting organic semiconductor with the work function of the electrodes.

2.2.1. *Meaning and advantage of aligning the energy levels*

The amplitude of the energy barrier depends on the energy levels of the materials on both sides of an interface. For an organic–organic interface, the energy levels in question are the HOMO and LUMO, and for an organic–electrode interface, it is the Fermi level which is concerned. To illustrate the concept of energy-level alignment, it is interesting to look at the movement of an electron, from its injection by the cathode up until its recombination with a hole.

When the electron is injected, it jumps from the Fermi level of the cathode (see Figure 2.2) to the LUMO of the electron transport layer (or the injection layer, if used). The electron then travels through the organic layer by jumping between the LUMO levels of neighboring molecules. Note that the movement of an electron within an organic layer is not exempt from an energy barrier, because when a charge carrier (an electron or hole) is placed on a molecule, the carrier polarizes the molecule and its immediate environment. Under the influence of that polarization field, the charge carrier is similar to a polaron. Once again, as it moves, the polaron has to overcome the potential barrier associated with the potential well left by the carrier, following the deformation of the lattice it has induced.

When the electron reaches the interface with a neighboring layer, it either jumps to the LUMO of that layer, crossing the energy barrier between the LUMO energy levels of the two materials, or recombines with a hole in the HOMO in the next layer.

Much like the electron, the hole is injected from the Fermi level of the anode into the HOMO of the hole transport material (or hole injector, if used). The energy barrier to the injection of holes is the difference between the Fermi level and the energy of the HOMO, represented as $\Delta\Phi a$ in Figure 2.2. Once injected, the hole moves through the materials, jumping between the HOMOs of neighboring molecules or layers.

These energy barriers to the injection of holes and/or electrons can be overcome by the application of an external electrical field. The minimal injection barriers are obtained when the molecular orbitals of the charge-injection organic semiconductors (LUMO and HOMO) are aligned with the Fermi levels of the electrodes used. On the side of the cathode, the LUMO of the electron-injection material must be aligned with the Fermi level of the cathode, and at the anode, the HOMO must be aligned with the anode's Fermi level. This phenomenon is known as energy-level alignment. Different possible mechanisms of injection at the interfaces can be envisaged, depending on the level of the energy barrier at the interface between the electrode and organic layer and on the electrical field applied to the OLED.

2.2.2. *The different mechanisms of charge injection at the electrodes*

Depending on the nature of the organic–metal junction, the height of the potential barrier at the interface and the value of the electrical field applied $E_a = V_{AC}/d$, where d is the thickness of the organic film between the two electrodes, three types of charge injection may exist [MOL 03]:

– thermoelectronic injection, where $E_a = 0$;

– field-effect injection (Schottky emission effect), with E_a being "of medium intensity";

– tunnel injection.

We shall also see a model that is frequently used to represent OLEDs: Scott and Malliaras' model [MAL 99].

2.2.2.1. *Thermoelectronic injection (T ≠ 0; Ea = 0)*

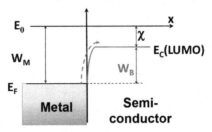

Figure 2.4. *Thermoelectronic injection*

First of all, let us look at thermoelectronic injection, induced by thermal agitation, which causes vibrational agitation of the molecules. This phenomenon was first described by Guthrie in 1873. Occasionally, the thermal energy manages to surpass the electrode work function, thereby producing a weak current. In 1901, Richardson's equation was written, to introduce the current density J linked to this phenomenon [RIC 01]:

$$J\left(W_B,T\right) = A_G T^2 \exp\left(\frac{-W_B}{k_B T}\right) \qquad [2.2]$$

where $A_G = \lambda_R \dfrac{4\pi k^2 e}{h^2}$

A_G: Richardson's constant

T: temperature in Kelvin

χ: electron affinity

W_M: metal work function

W_B: potential barrier, $W_B = W_M - \chi$

k_B: Boltzmann's constant

λ_R: coefficient dependent on the material

e: charge of an electron

h: Planck's constant

However, this model does not take into account the influence of the electrical field E resulting from the potential difference.

2.2.2.2. Field-effect injection (Schottky emission): E_a of medium intensity

The application of an electrical field E_a lowers the potential barrier between the metal and the organic semiconductor by ΔW, thus facilitating the injection of electrons, such that:

$$\Delta W = \sqrt{\frac{e^3 E}{4 \pi \varepsilon_0}} \qquad [2.3]$$

In order to represent the impact of the field E_a on injection, Schottky supplemented equation [2.2] as follows [KLE 12]:

$$J(W,T,E) = A_G T^2 \exp\left(\frac{-(W_B - \Delta W)}{k_B T}\right) \qquad [2.4]$$

To determine whether the current is limited by field-effect injection, we simply need to check whether the curve $\log(J) = f\left(\sqrt{E_a}\right)$ at constant temperature is linear.

In reality, this model needs to be considered with a degree of caution. The main problem is that Richardson's constant, extracted to adapt the OLED's electrical behavior, often assumes unrealistic values. In addition, as the model depends on the temperature and the electric field, further problems may manifest themselves, and the expected dependency cannot be verified experimentally. Naturally, this model is mainly useful in describing the fundamental physics of the interface between the metal and the organic semiconductor, which gives us an initial idea of the macroscopic behavior we can expect. In addition, it represents a good starting point for further improvements, taking account of the peculiarities of organic semiconductors.

In general, the model is reasonably well suited to the use of a relatively weak electrical field. Equation [2.4] becomes inaccurate when a strong field E_a is applied – typically greater than 10^8V/m. Indeed, a strong field E_a can give rise to injections by the tunnel effect.

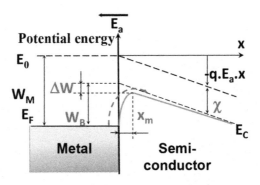

Figure 2.5. *Schottky injection*

Note that in Figure 2.5, x_m is the width of the space charge zone (depletion zone), which is due to the injection of charge carriers into the organic semiconductor, and it is given by:

$$x_m = \left(\frac{q}{16\pi\varepsilon E_a} \right)^{\frac{1}{2}}$$

[2.5]

2.2.2.3. Tunnel injection

When E_a is very intense, the potential energy curves become highly oblique, because the width of the depletion zone becomes very low (see equation [2.5]). Thus, the potential barrier encountered by the electrons becomes very thin: it can be crossed by the process corresponding to a tunnel effect (see Figure 2.6). The current density is given by the Fowler–Nordheim equation [FOW 28], such that:

$$J_{FN} = \frac{A^* e^2 E^2}{W_B C^2 k_B^2} \exp\left(\frac{2\,C\,W_B^{1.5}}{3\,e\,E} \right) \propto E_a^{\,2} \cdot e^{-\frac{k}{E_a}}$$

[2.6]

To find out whether the current is limited by tunnel injection, we merely need to check the linear behavior of the curve produced by:

$$\log\left(\frac{J}{E_a^{\,2}}\right) = f\left(\frac{1}{E_a}\right) \qquad\qquad [2.7]$$

The slope of this curve gives us the value of the potential barrier W_B, with

$$C = \frac{4\,\pi\sqrt{2\,m^*}}{h} \qquad\qquad [2.8]$$

m^*: effective mass of charge carrier

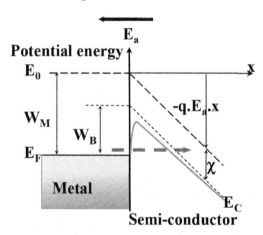

Figure 2.6. *Energy diagram of the metal/organic interface. Long dashes: injection of charges from the metal into the organic material (Richardson & Schottky). Short dashes: tunnel injection (Fowler & Nordheim)*

In the presence of a very powerful field E_a, the majority of injection takes place by the tunnel effect.

2.2.3. *Optimization of charge injection*

At the anode/organic material and cathode/organic material interfaces, the potential barriers are not the same. The potential that the charge carriers need to overcome in order to be injected into the organic material conditions the phenomenon of injection. It is important to choose metals whose work function is appropriate, enabling the charges to be injected without difficulty. The work function of the anode must be near to the energy level of the HOMO band of the organic layer receiving the electron holes, and the cathode work function must be close to the LUMO energy level of the layer receiving the electrons.

The efficiency of these steps depends on the properties of the materials: suitability with the work function of the electrodes at the energy levels of the carriers, appropriate carrier mobilities, establishment of sufficient concentrations of electrons and holes in the right places in the diode, favorable for light emission and minimizing non-radiative recombinations.

2.2.3.1. Choice of anode

To minimize the barrier $\Delta\Phi_a$, the work function of the anode must be chosen to be as near as possible to the HOMO energy level of the organic material used (generally near to 5 eV). In addition, the anode must be transparent, to allow the emitted light to shine out. The majority of OLEDs use metal oxides as anode contacts [GAR 96] – specifically, indium tin oxide (ITO), whose work function is near to 4.8 eV.

Karasawa and Miyata [KAR 93] showed that ITO has a resistivity close to 7×10^{-4} Ω.cm, which corresponds to a good electrical conductor. The transmittance of this material depends slightly on the thickness of the layer deposited. However, it remains greater than 70% across the whole visible spectrum (380–780 nm), with a value of 90% for a thickness of 420 nm.

The work function of ITO depends on its stoichiometry, on the deposition conditions and on the surface treatment. The HOMO of the organic layers is generally lower than the Fermi level of ITO (typically 5–6 eV, as compared to 4.5–4.8 eV). The interface barrier standing in the way of the injection of the holes therefore must be minimized as far as possible. Multiple surface treatment methods are used to increase the work function of ITO. Wu et al. [WU 97] showed that by treating ITO with a gaseous plasma, it was possible to alter the optoelectronic characteristics of the OLEDs without modifying the properties of the ITO (square resistance and transmittance). They observed that oxygen plasma treatment improves the quantum yield of the diodes (1% instead of 0.28%) and significantly decreases the threshold voltage (3 V instead of 12 V).

Kugler *et al.* [KUG 97] demonstrated that different treatments could modify the value of ITO's work function:

- 4.5 eV with a conventional acetone-ethanol treatment;

- 4.8 eV with oxygenated water treatment (H_2O_2);

- 4.0 eV with a negative ion plasma.

A different approach to improve hole injection is to insert a very fine layer between the ITO and the organic layers. For example, Deng *et al.* [DEN 99] use a fine layer of SiO_2 to improve the balance between holes and electrons in the NPB/Alq3 bilayer structure whilst improving the adhesion of the NPB to the ITO. On similar structures (TPD/Alq3), Jiang *et al.* [JIA 00] use a thin layer of Si_3N_4 to treble their light yield (from 0.4 lm/W without a buffer layer to 1.2 lm/W with a buffer layer 2 nm thick). In all cases, this fine buffer layer improves the ITO/organic layer interface by limiting the diffusion of the ions in the organic layer and decreasing the number of sites of non-radiative recombinations at the interface.

2.2.3.2. *Choice of cathode*

As in the case of the anode, the work function of the cathode is a crucial parameter. The LUMO levels for the majority of organic materials used are between 2.5 and 3.5 eV. To facilitate the injection of electrons from the cathode, that cathode must have a low work function. The metals with the lowest work function are alkalis (1.8 eV for cesium and 2.3 eV for sodium) and, particularly, lithium (2.3 eV). However, lithium cannot be used, as its extremely low melting point and its instability make it difficult to evaporate in its simple metallic form.

Making cathodes out of calcium, aluminum or magnesium alloys has yielded excellent results. However, these materials are vulnerable to air and humidity, which drastically degrade their electronic properties. Many research groups are presently focusing on the development and optimization of encapsulation materials, designed to protect the organic structures against such attacks.

In order to reduce the Schottky energy barrier at the cathode/organic material interface and encourage tunnel injection of charges, numerous teams have inserted a fine layer (0.1 nm–1.5 nm) of LiF [BRA 02], CsF [PIR 00, JAB 00] or Li [PAR 01] between the organic material and the cathode. Although the results are conclusive, the mechanism to lower the barrier has yet to be clarified; the action of lithium is complex, and the reaction mechanisms appear different on Alq3 [MAS 01] and on polymers [BRO 00].

2.3. Charge transport

Charges with opposite signs, coming from the two electrodes, move through the injection layer and charge transport layer, before meeting and recombining in an emission zone, ideally situated in the middle of the OLED heterostructure. The designation of a material as an "electron transporter" or a "hole transporter" is not determined only by the difference between electron mobility and hole mobility. It is also governed by the material's capacity to inject a certain type of charge from conductive electrodes or underlying layers and transport them to the emissive layer. In addition to a high charge mobility, this transport layer must have an adequate level of energy for the charges to be able to move toward the emissive layer without needing to cross high energy barriers. For example, NPB (N,N'-Di(1-naphthyl)-N,N'-diphenyl-(1,1'-biphenyl)-4,4'-diamine) is often used as a hole transport material, despite the fact that the mobility of electrons is higher than that of holes.

2.3.1. *Hole transport layer*

As mentioned previously, a high-performing OLED is made up of multiple organic layers, with the gaps in energy levels between the layers acting as energy barriers against the transport of charges to the emissive layer. For a hole transport layer (HTL), its HOMO level determines the charge injection barrier from the anode and from the anode to the emissive layer, while a sufficiently high LUMO level offers the electron blocking necessary to confine the charges in the emissive layer. There are other requirements as well: chemical stability of the HTLs while the device is in operation, high conductivity to ensure low operating voltage and an appropriate triplet level which ensures the excitons are confined within the emissive layer (EML) and prevents their quenching at the EML/HTL interface. The development of an HTL material which simultaneously satisfies all these criteria remains a subject of study, and a major challenge.

An effective strategy, therefore, is to obtain the desired behaviors and functions by combining the properties and advantages of multiple HTLs.

In a high-performing OLED heterostructure, we usually find two hole transport layers, and less frequently a third electron blocking layer. The first layer is known as a Hole Injection Layer (HIL). Materials with low ionization potentials are used as the first HTL (or HIL) to facilitate the injection of holes from the anode into the organic layers. This layer must, obviously, exhibit good hole mobility to facilitate their transport. For example, 4,48, 49-trishN,s3-methylphenyld-Nphenylaminoj-triphenylamine) (m-MTDATA) and its family – e.g. 4,4', 4"-tris[1-, 2-naphthyl (phenyl)amino] triphenylamine (1 and 2 TNATA) [SHI 97] and 4,4', 4"-tris [9,9-dimethylfluoreno-2-yl (phenyl) amino] triphenylamine (TFATA), are characterized by their very low ionization potential and amorphous films in high-quality thin layers. Shih-Fang Chen [CHE 04] presented a detailed study of the influence of HILs on the performances of Alq3-based OLED heterostructures. The band diagram of these OLEDs is shown in Figure 2.7, using N,N8-diphenyl-N, N8-biss1-naphthyl-phenyld-s1,18-biphenyld- 4, 48-diamine (NPB) as a hole transport layer.

Figures 2.8(a) and 2.8(b) illustrate the current–voltage and luminance–voltage characteristics for these various voltage OLEDs [CHE 04]. The results show that the two characteristics J-V and L-J of OLEDs were significantly improved when a hole injection layer (NPB) was inserted between the ITO anode and the HTL. In addition, of the various hole injection materials used, the OLED containing a layer of m-MTDATA exhibits much better performances than other OLEDs. The insertion of the m-MTDATA layer enabled holes to be more efficiently injected and conducted into the emissive layer. The improvement of luminous efficiency shows better charge balances between holes and electrons when such a HIL is inserted.

An injection layer must be chosen on the basis of three criteria:

– the value of the hole mobility: it is very often necessary, for a variety of reasons, to deposit a thick HIL. In these conditions, a high charge mobility is necessary to maintain a low operating voltage;

– the HOMO: in a hole injection layer, we want to have a HOMO level which is intermediate between the work function of the anode and the HOMO of the HTL;

– the surface morphology: an injection layer must be able to provide a flat surface, with very low roughness. This helps create a uniform conduction and percolation path for the holes injected from the cathode to the HTL. An injection layer is often used to smooth the surface of an ITO substrate; such a mechanism could be a dominant factor, which leads to a decrease in the leakage current in the OLED, and thus increases its luminous efficiency. The m-MTDATA layer is well known for its smoothness, which encourages homogeneous and effective migration of holes from the m-MTDATA layer to the HTL.

Other phenomena may also play a crucial role in the choice of injection layer, such as the levels of traps and their activation energies [STA 99].

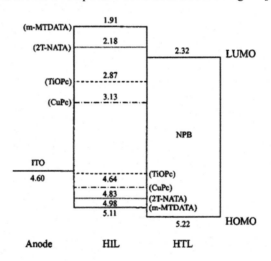

Figure 2.7. *Band diagram for different ITO/HIL/HTL structures [CHE 04]*

CuPc layers have also been widely used as a hole injection layer [LEE 15]. We also very frequently find OLEDs using very stable polymer films based on poly(3,4-ethylenedioxythiophene):poly(styrene sulfonate) (PEDOT:PSS), which has relatively high mobility.

To facilitate electron transport to the emissive layer, we use a second hole transport layer, which is characterized firstly by a HOMO energy level that is intermediary between that of the HIL and the emissive layer, and secondly by a relatively high hole mobility.

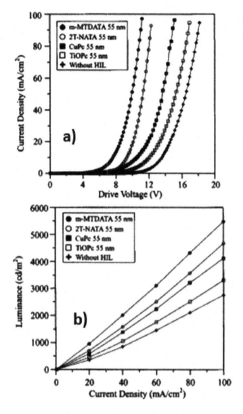

Figure 2.8. *Characteristics (a) J-V and (b) L-J of Alq3-based OLEDs with different hole transport layers [CHE 04]*

In OLEDs made with small molecules, triphenylamine-based materials having good charge mobility are usually used [CIA 11]. By way of example, TPD is a well-known hole transporter thanks to its level of ionization, its mobility and chemical stability. However, its glass transition temperature (T_g) is very low (around 65°C), leading to problems of crystallization and loss of performance when at high excitation in an OLED. This problem has been partially regulated with another derivative, which is α-NPD, which exhibits a glass transition temperature of around 95°C. Other molecules, based on carbazole (e.g. TCTA) are also used to transport holes.

In spite of their relatively high transition temperature (150°C for TCTA), their charge mobility is much lower than that of α-NPD. Time-of-flight (TOF) measurements of mobility have shown that the mobility of TCTA is of the order of 3×10^{-4} cm²/Vs [KAN 07], whilst that of α-NPD is approximately 2.2×10^{-3} cm²/Vs [AON 07].

Experimental case studies to demonstrate the advantage of introducing the two HTLs were carried out on a series of OLEDs, varying the thicknesses of the layers of α-NPB and m-MTDATA. The results of characterization of these OLEDs are displayed in Figure 2.9.

Figure 2.9(a) shows the I-V characteristics, which indicate that the insertion of two HTLs significantly increases the density of charges injected into the heterostructure. Following the addition of the layer of α-NPD, the voltage decreased from 16 V to 12.5 V, and the current density trebled to 20 V. This is attributable to the fact that the potential barrier at the m-MTDATA/Alq3 interface is too great, and that it is necessary to have a strong field for the holes to be able to cross that barrier. α-NPB ensures the transport of holes to the emissive layer thanks to its HOMO energy level which is intermediary between the HOMO level of the m-MTDATA and that of the Alq3, as shown by the energy diagram in Figure 2.9(a). Indeed, in Figure 2.9(b), we can see that the absence of α-NPB in the heterostructure decreases the luminance by a factor of 4. This is due to poor charge injection at the m-MTDATA/Alq3 interface due to too high an energy barrier.

A material such as α-NPD satisfies two important criteria: it has high hole mobility and good long-term stability when the device operates.

Thus, α-NPD is a very viable choice to make OLEDs, having a matrix which emits in the green or red spectral regions. However, its low triplet energy (2.3 eV) means it is a poor candidate for use as a hole transport layer underlying the matrices containing blue phosphorescents, due to the quenching of the matrix's triplet states by the triplet states of the α-NPD layer, thus necessitating the intercalation of a fine exciton blocking layer between the α-NPD layer and the emissive layer. Apart from a high triplet level, this exciton blocking layer must also have good thermal and chemical stability and an adequate HOMO level to allow the holes to pass into the emissive layer.

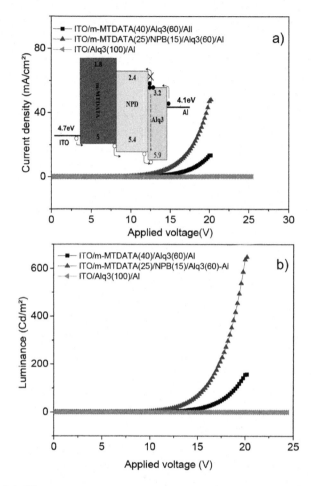

Figure 2.9. *The current density a) and luminance b) of different Alq3-based OLEDs. For a color version of this figure, see www.iste.co.uk/boudrioua/lasers.zip*

However, high hole mobility is not entirely necessary, because usually the exciton blocking layer is thin. For example, Tris(4-carbazoyl-9-ylphenyl)amine (TCTA) is generally chosen to block excitons at the HTL/ETL interface in OLEDs containing a blue phosphorescent. Although its hole mobility is much lower than that of α-NPB, a layer of TCTA less than 10nm thick represents no handicap to the performances of the OLED.

This thickness of 10 nm is sufficient to confine the excitons generated in the blue emission zone. Note that the stability of the TCTA layer during the OLED's operation raises a number of concerns [SIV 09]. Thus, the search for an HTL material which can serve all the requirements mentioned above is also an important field of research.

	Hole mobility (cm²/V.s)	LUMO	HOMO	E_t (triplet energy)	Tg
CuPC [KIT 04]	1×10^{-3}	3.6	5.3	–	>200
TPD [NAY 10]	2×10^{-3}	2.2	5.56	–	65
α-NPD [AON 07]	0.18×10^{-3}	2.4	5.4	–	96
TCTA [PAR 09]	2.0×10^{-5}	2.3	5.7	2.7	151
Pedot-PSS [RUT 13]	12.8×10^{-3}	2.4	5.2	–	–
m-MTDATA	0.15×10^{-3}	1.9	5.1	2.4	75
CBP [KIM 07]	–	2.5	5.8	2.6	–

Table 2.1. *Hole mobility, HOMO, LUMO, triplet energy level and glass transition temperature of different hole transport materials*

2.3.2. *Electron transport layer*

The Electron Transport Layer (ETL) must also simultaneously possess multiple properties: an adequate LUMO energy level to facilitate the injection of electrons on both sides of that layer, high electron mobility to facilitate their transport, and a high HOMO level, to block the holes and excitons.

Owing to the difficulty of simultaneously satisfying all the necessary criteria, a much lower number of ETLs, in comparison to the number of HTLs, have been reported to show efficient use in OLEDs. In particular, the mobility of electrons is generally lower in organic materials than that of holes. In addition, the chemical stability in a working OLED is more difficult to achieve for an electron-transporting material than for a hole-transporting material.

The first OLED using an ETL was developed by Adachi [ADA 88, ADA 90], who used a layer of TPD for hole transport and a layer of perylene bis-benzimidazole (PV) as an electron transport layer (see Figure 2.10).

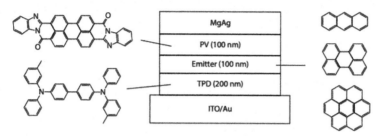

Figure 2.10. *A multilayer structure proposed by Adachi, using a hole transport layer and an electron transport layer [ADA 88]*

The results obtained have shown good efficiency of double charge injection by the use of a transport layer for each type of charge. The HTL and ETL proposed have also been able to block the electrons and holes, respectively, to confine them to the emissive layer. This device is the basic prototype for all multilayer OLEDs.

At present, in the case of green and red OLEDs, metallic complexes – Alq3 is one of several possible examples – are widely used as electron transport materials. However, these materials cannot deliver a sufficient level of electron blocking to create blue OLEDs. For example, when we use materials having properties which favor the transport of holes, such as TPD or α-NPD, as emitters in the blue or blue-violet, rather than as hole transporters, the injection of holes from these materials into the Alq3 layer gives rise to emission in the green spectrum. An approach similar to that put forward for the hole transport layers can be suggested to separate the two functions, of electron transport and hole blocking. This approach consists of using at least two layers: the first is a high electron-mobility layer to facilitate the transport of the electrons from the cathode to the emissive layer; the second is a fine layer to block holes in the emissive layer, and is intercalated between this latter and the electron transport layer (see Figure 2.11).

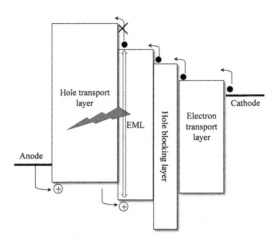

Figure 2.11. *OLED heterostructure with an
electron transport layer and a hole blocking layer*

The materials used to make electron blocking layers must conform to various constraints. They must exhibit an acceptable electron mobility, a high ionization potential (HOMO level), and must be able to accept electrons from the hole transport layer and inject them into the emissive layer. In other words, the LUMO of the electron blocking layer must be much higher than that of the emissive layer, and the HOMO level of the blocking layer close to that of the emissive layer. This last constraint is also necessary to avoid the formation of exciplex-type excitons at the interface between the emissive layer and the confinement layer. An exciplex is an exciton formed at the interface between two organic materials and involving two charges belonging to two different materials.

Numerous studies on OLEDs introduce a layer of BCP (2,9-dimethyl-4,7-diphenyl-1,10-phenanthroline) (Figure 2.12) into the heterostructure [HWA 05, NAR 13]. This material has the peculiarity of having a LUMO of the order of $E = 3.2$ eV and a significantly lower HOMO ($E = 6.7$ eV), which confines the holes within the emissive layer. However, the use of the BCP can lead to an increase in the working voltage of the OLEDs because of its low electron mobility. In addition, BCP tends to form exciplexes with numerous emissive materials having hole transport properties, such as TPD, α-NPD and CBP producing a red-shifted emission.

3-(4-biphenylyl)-4-phenyl-5-(4-tert-butylphenyl)-1,2,4-triazole (TAZ) is a well-known large-gap material with good electron mobility and chemical stability (Figure 2.12).

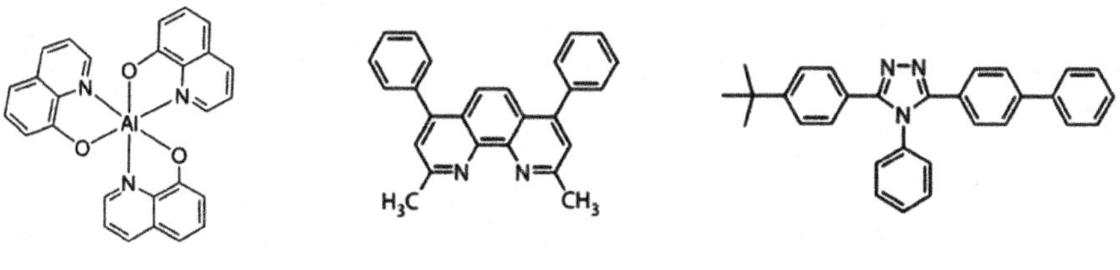

Alq3   TAZ   BCP

Figure 2.12. *Molecular structures of Alq3, TAZ and BCP*

When designing an OLED heterostructure, particular attention must also be paid to the nature of the emissive layer used and the type of charges it conducts best. In addition, when a new emissive material is first tested, many preliminary tests must be carried out to obtain as much information possible about its optical absorption and its emission, as well as its energy levels. Yet it is only once we have compiled precise data about the emissive layers that it is possible to evaluate the most compatible OLED structure. This structure must take account of the following factors:

– the confinement of charge carriers must be completely optimized in the emissive layer;

– the charge injection barriers must be low (as low as possible). This can be achieved by using an appropriate electrode and charge injection layer;

– the charge transport layers (ETL and HTL) must also, if possible, play the role of charge blocking (holes and electrons, respectively);

– to increase the confinement of the charge carriers, a hole blocking layer (between the EL and ETL) and/or a hole blocking layer (between the EM and HTL) can be used;

– the thickness of the different layers of the OLED heterostructure must be carefully studied: it is difficult to obtain a thin layer with good uniformity, but a thick layer can significantly increase the working voltage.

2.4. Recombinations of charges and generation of excitons

The electrons injected from the cathode and the holes injected from the anode are transported to and within the emissive layer. The charge carriers attract one another by Coulomb force and form electron–hole pairs (excitons) on a single molecule when the distance between two charge carriers is less than the Coulomb capture radius (r_c):

$$r_c = \frac{e^2}{4\pi\varepsilon_0\varepsilon_r k_B T} \tag{2.9}$$

At ambient temperature (T = 300 K) for ε_r = 3 (typical value of organic materials), the capture radius is $r_c \approx 19$ nm.

Generally speaking, the recombination process is explained by the Langevin theory and the recombination rate is [WET 11]:

$$\kappa_L = \frac{e(\mu_e + \mu_h)}{\varepsilon_0\varepsilon_r} \tag{2.10}$$

μ_e: electron mobility

μ_h: hole mobility

The recombination of the electrons and holes is a process that is independent of their spins, giving rise to the following four possible states: $|\uparrow\uparrow\rangle$ (triplet), $|\downarrow\downarrow\rangle$ (triplet), $\frac{1}{\sqrt{2}}(|\uparrow\downarrow\rangle + |\downarrow\uparrow\rangle)$ (triplet) and $\frac{1}{\sqrt{2}}(|\uparrow\downarrow\rangle - |\downarrow\uparrow\rangle)$ (singlet). The likelihood of creating a triplet by the recombination of an electron and a hole, therefore, is 75% – i.e. three times greater than the likelihood of creating a singlet. Consequently, the maximum internal quantum efficiency of devices using fluorescent materials is theoretically limited to 25%.

The efficiency of radiative de-excitation of the excited states depends primarily on the emission system used. One of the most important discoveries and innovations in the field of OLEDs was the use of matrix–dopant or "Host–Guest" systems. In addition to using fluorescent dopants, this system has enabled us to use phosphorescent emitters, and thereby attain record levels of efficiency [KAW 05]. OLEDs using a host–guest system with an internal efficiency of around 100% have been reported, using a phosphorescent material. In a host–guest device, excitons are formed in the host matrix material, and then the energy is transferred to the receiving molecule (the guest, or dopant) by a non-radiative process.

2.4.1. *The ideal host material*

In phosphorescent-based OLEDs, an internal efficiency of 100% is theoretically achievable, thanks to the radiative emission of singlet and triplet excitons. The process of Förster and/or Dexter energy transfer between the host and the guest, as discussed in Chapter 1 of this book, plays a crucial role in the confinement of the excitons in the dopant [TAN 08, SUN 06].

An efficient host layer must satisfy a number of constraints – namely:

– high charge mobility;

– an adequate LUMO level to accommodate the charges from the ETL, and a HOMO level that is deeper than the HTL;

– for efficient energy transfer, the two singlet and triplet states of the host matrix must be much higher than the excited triplet state of the guest molecule [AND 01]. In practice, this means that a triplet level ≥ 2.75 eV is necessary to ensure efficient energy transfer and prevent energy transfer in the opposite direction (i.e. from the guest to the host material).

Host materials can be classified on the basis of the width of their gap band, and of the energy of their triplet states; small-gap ETL hosts are used to make green or red OLEDs, and large-gap HTL hosts are used to generate emission in the blue spectral region or to make white OLEDs.

2.4.2. *Electron-transporting host materials*

The most widely studied electron-transport host material is indubitably Alq3, which was the first host material used in an OLED, by Tang *et al.* [TAN 87]. It has a HOMO level of 5.7 eV, and a LUMO level of 3.0 eV. The triplet energy level of Alq3 is relatively low: 2.0 eV. Its fluorescence spectrum is centered at 530 nm; therefore, it is an ideal host for red emitter guests.

There is also 1,3,5-tris (N-fenilbencimidizol-2-yl) benzene (TPBi) (Figure 2.13), which is a well-known ETL host [YAN 12]. It has a HOMO level of 6.2 eV and a LUMO level of 2.7 eV (so the gap band is 3.5 eV wide). Given the high value of its gap and its HOMO and LUMO levels, this host is very well suited for fluorescent and phosphorescent guest materials emitting in green and red. It is also suitable for certain blue guests [SHI 02b]. TPBi is also used as a hole blocker, in view of its high HOMO level.

Another well-known host material is 3-phenyl-4- (1'-naphthyl) -5-phenyl-1,2,4-triazole (TAZ) (Figure 2.13), which has a HOMO level of 6.6 eV and a LUMO of 2.6 eV. This material is often used to make green phosphorescent OLEDs based on Tris[2-phenylpyridinato-C2,N]iridium(III) (Ir(ppy)3).

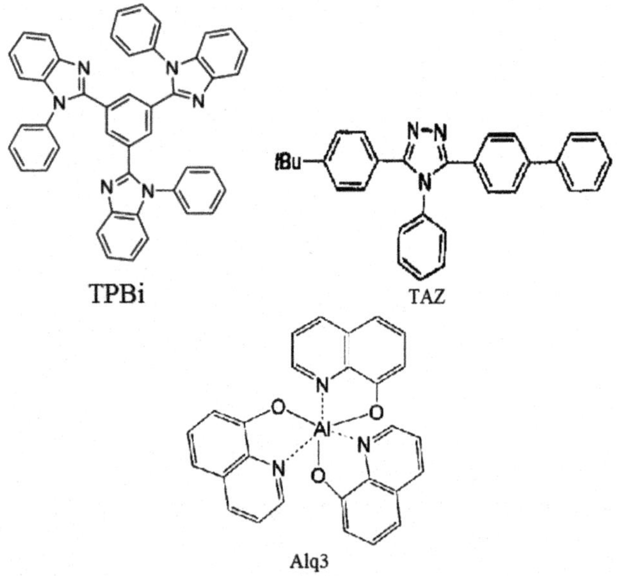

TPBi   TAZ

Alq3

Figure 2.13. *A number of electron-transporting host materials*

2.4.3. *Hole-transport layer host materials*

4,4'-bis (9-carbazolyl) biphenyl (CBP) is a widely used material to host phosphorescent and fluorescent emitters. Its HOMO and LUMO levels are, respectively, 5.8 and 2.5 eV. The triplet energy level is 2.6 eV. It has bipolar transport characteristics [KAN 97] – a property meaning it can serve as a good host material for green, yellow and red fluorescent and phosphorescent emitters. It is also used as a host in OLEDs based on blue fluorescent guest materials [CHE 16]. However, its low triplet level means it is unable to efficiently host blue phosphorescent dopants. A study has been published on a blue phosphorescent OLED (phOLED) based on a CBP host doped with a FIrpic blue emitter (iridium (III) bis [(4, 6-difluorophenyl) pyridinolate-N, C2] picolinate) [TSU 07]. Unfortunately, the relatively high triplet level in FIrpic (2.75 eV) means there cannot be an efficient energy transfer from CBP to FIrpic.

Another carbazole derivative, N,N'-3,5-dicarbazolil benzene (mCP) is also used [KES 13]. CBP and mCP have similar charge-transport properties. However, mCP has a relatively high triplet level, of the order of 3 eV. This high triplet energy means there is efficient energy transfer from mCP to the blue phosphorescent. By using mCP as a host material and FIrpic as a phosphorescent dopant, external quantum yields of 7.5% have been achieved. This value is 50% higher than the yield of a similar OLED using CBP (where η_{ext} = 5%) as a host material.

Another CBP derivative, 4,4'-bis(9-carbazolyl)-2,2'-dimethyl-biphenyl (CDBP), has been shown to spectacularly improve the performances of blue and white phosphorescent OLEDs, indicating efficient energy transfer from the triplets to the dopants [TOK 03].

The efficiency of fluorescence is often calculated on the basis of the response of an emitter material to an optical excitation. When the material is optically excited, the main loss mechanisms are vibrational relaxations and inter-system crossing (ISC). However, in an OLED, the electrical excitation engenders multiple loss mechanisms, which can drastically reduce the efficiency of fluorescence. Indeed, when an electrical excitation is applied, the polarons and triplets are directly created, respectively, by the phenomena of injection and recombination of the charges, leading to bimolecular annihilations which diminish the efficiency of the emissive medium. For further details on the different loss mechanisms, readers can refer to Chapter 1.

2.5. Dopants (guests)

To be compatible with the host materials cited above, the basic requirements for guest materials are:

– high fluorescence efficiency;

– an effective energy transfer between the host and the guest materials; the HOMO and LUMO energy levels of the guest and host must be well adjusted and well suited. The LUMO and HOMO levels of the guest molecule must be less deep than those of the host material;

– for triplet emitters (phosphorescence), it is necessary that the triplet energy level is less high than the host material.

2.5.1. *Dopants emitting in the red spectrum*

Emission at high wavelengths (in the red) is generally obtained by increasing the length of the π conjugation. Indeed, the greater the conjugation length, the smaller the gap between HOMO and LUMO becomes. However, this increases the probability of π-π *stacking*, because of the overlap between the π molecular orbitals, which limits the material's emissive efficiency. These red emitters have poor emissivity, or even none at all, in the solid state. Therefore, the Guest–Host system becomes a necessity to solve the problem of these red emitters when applied to OLEDs. The dopant molecules are dispersed and isolated in the host materials, which act as spacing brackets in the molecular structure, helping to reduce non-radiative annihilations. However, the optimum concentration of dopant is usually quite low – typically less than 2%.

Depending on their molecular structure, we can essentially distinguish two types of red fluorescent dopant:

– *Molecules with a pyran nucleus*: although a great many red fluorescent dyes have been synthesized and studied, DCM-type dyes – particularly DCJTB (see Figure 2.14) – remain the most efficient materials of them all. DCM and DCM2 were used by Tang *et al.* from Kodak Laboratories, in the seminal report demonstrating the dopant approach in improving the performance of OLEDs. The maximum electroluminescence of OLEDs containing DCM or DCM2 depends heavily on the concentration, within the ranges of 570–620 nm and 610–650 nm, respectively.

– *Rare earth complexes*: electron transitions for rare earth metals take place between discrete levels, and this results in highly monochromatic light absorptions and emissions. Thus, numerous energy levels closely follow one another between near infrared (NIR) and ultraviolet (UV). To limit the problem due to the low absorption of lanthanide ions, the trivalent ions can be inserted into a complex, formed with ligands by single or double bonds to form chelates with a high absorption coefficient, which transfer energy to the central ion. The complexes thus formed are soluble in many known organic solvents, such as chloroform and chlorobenzene. Consequently, they are also easily dissolvable in polymer matrices, unlike pure lanthanide salts. To prevent the phenomenon of π-stacking, rare earth complexes are inserted into a host matrix, which transfers the energy to the complex after any excitation (be it optical or electrical). The complex then emits the luminescence which is characteristic of the central ion. The excitation is "captured" by the host matrix, which transfers it to the central ion by way of the energy transfer mechanism represented in Figure 2.15. The matrix is excited (by electroluminescence or photoluminescence), and a Förster-type energy transfer from a singlet state level of the host matrix to an excited level of the ligand occurs. A better yield in terms of energy transfer can be obtained when the energy difference between the two levels is slight. If that is not the case (the energy of the host material and the ligand are not well matched), no energy transfer takes place, and only the host material itself may emit light.

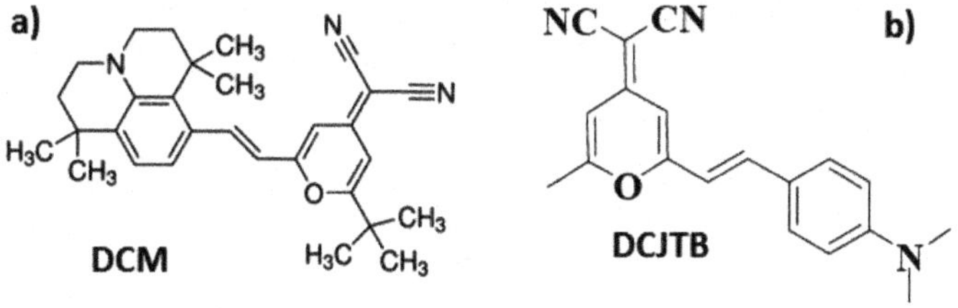

Figure 2.14. *Chemical diagrams of DCM and DCJTB*

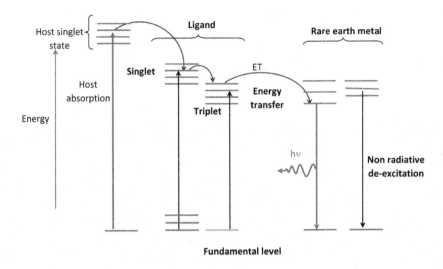

Figure 2.15. *Excitation and de-excitation of a rare earth metal molecule dispersed in a polymer host matrix*

Fang *et al.* [FAN 03], using an OLED based on Eu(Tmphen)(TTA)3, with an emission peak at 612 nm having a full width at half maximum (FWHM) of 3 nm, succeeded in obtaining a luminance of 800 cd/m², an external quantum yield of 4.3%, a luminous efficiency of 4.7 cd/A and an energy efficiency of 1.6 lm/W. However, the efficiency decreases when the current is increased, due to triplet–triplet annihilation in the ligands. Many other complexes such as Eu(TTA)3s phen (TTA = thenoyltrifluoroacetone) or Eu(DTP)3 (dipphen) (DTP = 1,3-di(2-thienyl)propane-1,3-dione, dipphen = 4,7-Diphenyl-1,10-phenanthroline) [CHA 06] have often been synthesized and inserted into red OLEDs.

In terms of emission, applications have been developed, in connection with the industrial availability of rare earth metals at sufficient levels of purity: color television, fluorescent lighting and medical radiography. A great variety of emission can be obtained as a function of the nature of the rare earth metal involved and the respective positions of the excited and fundamental energy levels. Although the purity of the color, chemical stability and efficiency are worthy of interest, they are as yet insufficient for the needs of commercial applications.

2.5.2. *Green-emitting dopants*

In the existing body of literature, we can distinguish several different types of green emitters:

– *Metal chelates*: most vapor-phase deposited OLEDs are made with tris 8-hydroxyquinolate aluminum (Alq3), because it has high electron transport capacity and advantageous fluorescence properties. Alq3 emits a very saturated green color (0.32, 0.55), and is, today, still one of the best green emitters available. Similar properties have been reported in many other chelate-metal complexes such as bis (2-2-hydroxyphenyl)benzoxazolate) $(Zn(BOX)_2$ and tris (4-phenanthridinolate) aluminum (Alph$_3$).

– *Coumarin-derived dopants*: of the fluorescent dopants, one of the best green dopants, used in numerous OLED heterostructures is 10-(2-benzothiazolyl)-1,1,7,7-tetramethyl-2,3,6,7-tetrahydro-1H, 5H, 11Hbenzo pyranno [6,7,8-ij] quinolizine-11-one (C-545T). This agent was used to create the first colored OLED host matrix, put on the market by Pioneer in 1998. C-545T and its derivatives, such as C-545TB and C-545mt, have high quantum efficiency of fluorescence in solution (up to 90%), and significant quantum yields when they are used in OLEDs [CHE 00].

– *Aminoanthracene derivatives*: these emissive materials have been used to make OLEDs with highly advantageous luminous yields, and very powerful emission in green. Yu *et al.* were able, using the emitter β-NPA, to obtain a maximum luminance in the green spectrum of around 65,000 cd/m², with a maximum quantum yield of 3.7% and a luminous efficiency of 14.8 cd/A [YU 02].

2.5.3. *Blue-emitting dopants*

In the case of blue emitters, the major challenge for the design of devices lies in obtaining stable blue emitters and hosts. There are a certain number of stable blue host materials which have been described in the literature. Broadly speaking, they can be classified into several major classes:

– *Diarylanthracenes*: these compounds have high quantum yields and emission colors centered firmly in the blue range, which makes them highly advantageous. In 2002, Shi and Tang at Kodak [SHI 02a] demonstrated a very stable OLED made with 2, 5, 8, 11- (-butyl)perylene (TBP), doped in

9, 10- (2-naphthyl) anthracene (ADN) as a blue emitter in a heterostructure of ITO / Copper(II) phthalocyanine (CuPc) (25 nm) / 4,4- [-(1-naphthyl)-phenylamino] biphenyl (NPB) 50 nm) / ADN: TBP (30 nm) / Tris-(8-hydroxyquinoline)aluminum) Alq3 (40 nm) / Mg: Ag (9: 1, 200 nm)]. The resulting OLED had a luminous efficiency of 3.5 cd/A, and chromatic coordinates of (0.15, 0.23). The OLED's reported working lifetime was around 4000 hours, with an initial luminance of 636 cd/m, which was one of the best results ever published at the time. However, more in-depth studies revealed a morphological instability of the ADN layer when the OLED is subjected to high current for a prolonged period of time.

In order to improve the morphological stability of the thin layers of ADN, various ADN derivatives, adding a butyl group, such as 2,6- (butyl) -9,10- (2-naphthyl) anthracene (DTBADN) and 2,6- (-butyl) -9,10- [6- (-butyl) -2-naphthyl] anthracene (TTBADN) have been synthesized and compared [LEE 05]. However, it has been shown that the addition of these groups causes a significant reduction in its quantum fluorescent efficiency. To improve the morphological stability of a thin layer of ADN without adversely affecting its emission efficiency, other derivatives have been put forward, including α- and β-TMADN [LEE 04a] and MADN [KAN 04]. With these two emitters, it is possible to achieve a luminance of 12,000 cd/m², maximum efficiency of 5.2 cd/A and emission centered at 466 nm and with CIE coordinates of (0.152, 0.229).

– Di(styryl)Arylenes: another host matrix that is well known in the OLED community is DPVBi (4,4'-bis(2,2-diphenylvinyl)-1,1'-biphenyl) (Figure 2.16), first used by Hosokawa and his colleagues at Idemitsu [HOS 95]. DPVBi has LUMO and HOMO levels of 2.8 eV and 5.9 eV, with a large gap of 3.1 eV. DPVBi has been able to deliver good performances in all OLEDs in which it has been used.

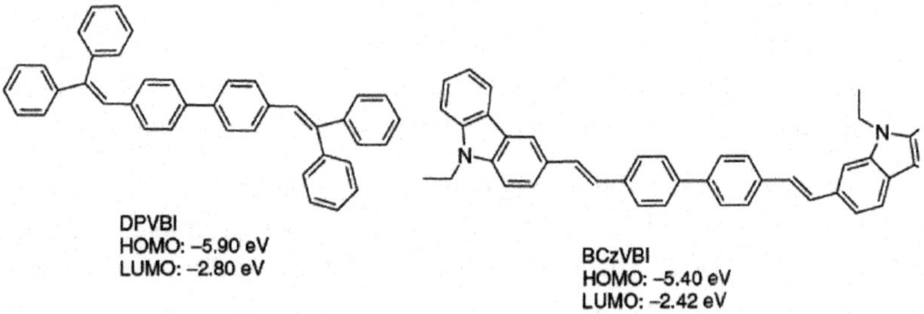

DPVBI
HOMO: –5.90 eV
LUMO: –2.80 eV

BCzVBI
HOMO: –5.40 eV
LUMO: –2.42 eV

Figure 2.16. *Chemical diagrams of DPVBi and BCzVBi*

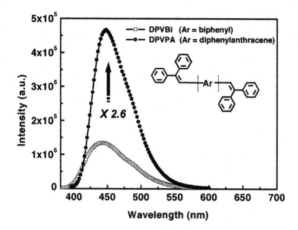

Figure 2.17. *Structures and PL spectra of DPVBi versus DPVPA [LEE 04b]*

However, in light of more recent studies, it has become increasingly clear that DPVBi could not serve as a very good blue matrix, because its fluorescence yield in solution is around 38%, which is much too low for efficient Förster energy transfer [WEN 05]. Another derivative – DPVPA (9,10-bis[4-(2,2-diphenylvinyl)phenyl]anthracene) – was proposed by Ali *et al.* [ALI 04], delivering an external quantum yield around 2.6 times greater than that obtained with DPVBi. However, the fluorescence spectrum in solution (in toluene) was centered at 448 nm, and was shifted around 20 nm toward green, because of the relatively longer conjugation of the DPVPA molecule [LEE 04] (see Figure 2.17).

– *Diphenylamino-di (styryl)Arylene (DSA-Ph)*: DSA-Ph constitutes the most efficient group of blue emitters known at present, and is used in the first generation of commercial blue OLEDs, both as an emitter and a host matrix. Hosokawa *et al.* [HOS 95] were the first to use an emitter from this family of materials: 4,4'-bis(9-ethyl-3-carbazovinylene)-1,1'-biphenyl (BCzVBI) (Figure 2.16). It was used as a dopant in a DPVBi matrix. The OLED structure created was as follows: ITO / CuPc / DPT / DPVBi:BCzVBI / Alq3 / Mg: AG. The maximum luminance was 10,000 cd/m², with an electroluminescence peak at 468 nm. The doped OLED delivered luminous efficiency of 1.5 lm/W and a maximum external quantum yield of 2.4%, which is around twice as great as that obtained with the same device, but without doping of the emissive layer with DPVBi. The DPVBi film had a good morphological stability, and the device had an estimated lifetime of over 5000 hours.

Ueno *et al.* [TSU 04, SUZ 04], after studying various host materials, noted that the use of 1,3,5-tris(1-pyrenyl)benzene (TPB3) as a host, and a dopant known as IDE-102 (Figure 2.18), produced an OLED capable of a luminosity of around 142,000 cd/m² at 12 V, a luminous efficiency of 6 lm/W and external quantum yield of 2.4% at 5 V.

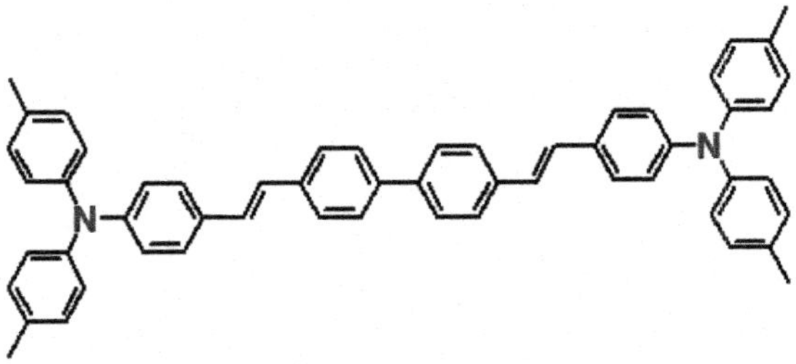

Figure 2.18. *Chemical diagram of IDE 102*

More recently, Chen *et al.* created a dark blue OLED based on DB1, whose emission is centered at 438 nm. This emitter has been used to obtain blue OLEDs (CIE coordinates: 0.14, 0.13), with very high yields (5.4 cd/A).

2.6. OLED fabrication techniques

Once we have chosen the organic materials we wish to use, it is also crucial to choose the most appropriate fabrication technique. Most organic materials are available in the form of powdered microgranules. However, there are several techniques available to deposit the organic layers which will make up the OLED. The majority of techniques have already been widely used and mastered in the domain of electronics or optical coatings.

Each technique brings its own particular advantages and disadvantages. As mentioned in Chapter 1, there are two choices for the organic layers: small molecules or polymers. The deposition method differs depending on which option is taken.

As previously indicated, ITO plays a crucial role in the operation and aging of OLED devices. It is therefore essential to have surface states as clean as possible. Indeed, when creating an OLED, we begin by chemically washing the substrates with the usual cleaning solvents – namely:

– ultrasonic acetone cleaning for 5 minutes;

– ultrasonic ethanol cleaning for 10 minutes;

– ultrasonic deionized water cleaning for 10 minutes;

– isopropanol for 5 minutes.

The clean substrates are then subjected to UV-ozone treatment for 15 minutes. The purpose of this step is to increase the work function of the ITO in order to lower the potential barrier which the holes have to cross.

2.6.1. *Vacuum deposition*

Vacuum deposition is the most widely used technique to deposit small organic molecules. It is a vapor-phase deposition process: the material being deposited is heated in a vacuum ($>10^7$ mbar) in a high-amp metal boat (see Figure 2.19). The high temperature first melts and then evaporates the organic powder (or else sublimates it, directly from the solid to the gaseous state). After evaporation, it condenses on the cooler parts of the apparatus and, in particular, on the surface of the substrate, thus forming a homogenous film. The number of crucibles in the evaporating system determines the total number of films that can be laid down without breaching the vacuum. The thickness of the film and rate of deposition are controlled *in situ* by a piezoelectric sensor, whose measuring principle is based on the modification of the frequency of a piezoelectric crystal by an overload in terms of mass ΔM of the crystal (see Figure 2.19). With this apparatus, it is possible to deposit extremely uniform, very thin layers measuring a few tens of Angströms, at a very slow rate of around 0.1 Angström/s.

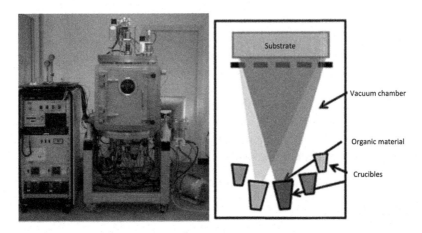

Figure 2.19. *Typical apparatus for deposition of organic layers to make an OLED. Principle of evaporation of the different materials in the chamber*

To prevent any contamination, the organic layers are laid down one after another, without being taken out of the chamber into the air. So as to have extremely stable, reproducible deposits, the evaporation of the organic materials requires a very long temperature ramp. We impose deposition rates of between 2 and 3 Å/s depending on the material being used. As with any other materials deposited by evaporation, too low a rate causes the inclusion of impurities in the film, whereas too high a rate can cause structural faults.

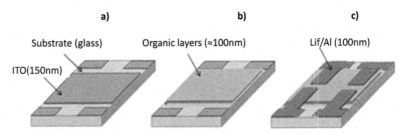

Figure 2.20. *a) Glass substrate/etched ITO. b) Deposition of the different organic layers. c) Final structure of the sample after deposition of the cathode (four OLEDs on the same sample)*

The different steps in the making of OLEDs are illustrated in Figure 2.20. It shows the glass substrate upon which a layer of ITO is deposited. The ITO is then etched to ensure the metal contacts and prevent a short-circuit occurring between the ITO anode and the Al cathode (a). Then, the organic layers are laid down, covering almost the entire surface of the sample (b). Finally, the last step is to deposit the metal cathode (generally Al) (c).

The advantages of vacuum deposition are:

– good control of the thickness and rate of deposition, thanks to the microbalance crystal quartz;

– possibility of creating extremely thin layers (~1 nm);

– possibility of making multilayered structures. The number of layers is limited by the number of crucibles (or boats) contained in the evaporation system;

– the thin layers are deposited at high vacuum, ensuring the purity of the deposit and preventing any contamination and oxidation of the organic layers by oxygen and humidity.

Despite the numerous advantages of the vacuum deposition technique, this method does have some not-insignificant drawbacks. Whilst it is simple in theory, vacuum deposition requires the use of a highly sophisticated, expensive piece of equipment, functioning at high vacuum. Furthermore, this technique has very low deposition efficiency: around 5% of the organic vapors are deposited on the substrate, and the rest on the cold walls of the vacuum chamber.

2.6.2. *Spin coating*

Spin coating has become the leading choice of most research centers in numerous domains, thanks to its simplicity, the relatively inexpensive equipment used, and the good results that can be obtained. It is method of centrifugal deposition (Figure 2.21). The organic material in solution is deposited on a substrate with a pipette. The substrate is fixed to a plate by an aspirator, and by spinning the plate, the material is made to spread uniformly across the whole surface of the substrate, due to centrifugal force. The rotation speed and acceleration of the spinning plate are the two main parameters controlling the thickness of the organic film. By steaming the sample, we are able to evaporate the coating solvent. This method is essentially used for deposition of polymers.

Figure 2.21. *Different stages of spin coating*

1) *Advantages to the technique of spin coating*:

– there is less loss of materials than with vapor-phase deposition;

– it is a relatively inexpensive technique – a spinning plate is far cheaper than a vacuum deposition system;

– the spin-coating means we can quickly and easily deposit thin layers.

2) *Disadvantages to the technique of spin coating*:

– difficulty of creating multilayer structures (> 2 layers);

– possibility of the presence of contaminants (traces of solvent, oxygen, humidity, etc.);

– difficulty of accurately controlling the deposition (homogeneity, rugosity, etc.);

– impossibility of creating extremely thin films (< 10nm).

2.6.3. Inkjet printing deposition

Since the late 1990s, the possibility of OLED deposition by inkjet printing has been touched upon and explored by numerous groups. The objective is to provide a low-cost method of manufacturing large-dimension OLED screens with higher resolution. The principle of inkjet deposition is based on printing on a substrate with tiny droplets (often measuring less than µl) of organic material in solution (ink drops) in specific positions or at specific points (x, y) – see Figure 2.22. The position of each deposition point is defined by the user of a specific control software program. Thus, it stands to reason that it is of interest in manufacturing color screens where the pixels (the points of deposition) must be perfectly aligned. In this case, multiple printing heads are used to handle the different organic materials, emitting the red, green and blue components of an RGB system. The important part of this technology is the printing head, which is specially developed to apply miniscule quantities of organic semi-conductors dissolved in specific solvents.

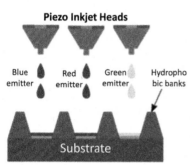

Figure 2.22. *Left: diagram of inkjet printing deposition. Right: an inkjet printing deposition device developed by Kateeva in the USA*

1) *Advantage to inkjet deposition*

This technology has the advantage of being easy to implement, and of being usable in normal atmospheric conditions. It can be used on almost any type of substrate (equally for glass substrates and flexible substrates, or paper). It should, in principle, enable us to carry out a deposition with a very high degree of accuracy.

However, to obtain the required resolution, processes need to be developed to determine the exact quantities of dissolved polymer which the printer needs to release at each position, and to ensure uniform printing whatever the type and thickness of the substrate and the type of polymer used.

The potential offered by this technology has already attracted the attention of major enterprises specializing in the domain of display.

On the springboard of a technology developed by MIT (Massachusetts Institute of Technology, USA), a startup company, "Kateeva", based in the US, was set up in 2009 to develop equipment for inkjet OLED deposition. It is now the best-known company in the area of inkjet printing, thanks to the YIELDjet system, designed for mass production of OLEDs on large substrates.

Working closely with the company Merck (specializing in synthesis of organic materials), in 2014, AU Optoronix developed an inkjet printer to make 14-inch OLED screens.

In March 2014, Tokyo Electron (TEL) announced the development of a printer called Elius 250. In 2010, TEL had begun working with Seiko Epson on OLED manufacturing technology, which would combine the inkjet printing method that is Epson's area of expertise with the OLED screen production apparatus developed by TEL.

Other companies have also shown an interest in inkjet deposition technology, such as LG Display and Konica Minolta. The latter, in 2012, developed a high-precision inkjet printing head capable of producing extremely small drops of 1 picoliter. The new printing head was the first to use the MEMS technology (Micro-Electro-Mechanical Systems) developed by the company.

2) *Disadvantage to inkjet deposition*

At present, there are a number of problems associated with the technique of inkjet deposition, meaning that this technology is somewhat scarce on the market, in spite of its huge potential.

The first is the problem of the uniformity of deposition of a polymer when deposited in a tiny quantity (pico-liters). The surface potential of the substrate tends to distribute the droplets and cause them to slip, which has dire consequences for the quality and uniformity of deposition. This is known as the "coffee stain effect", resulting from the accumulation of product at the edges of the ink droplets during the drying process [DEE 00].

The second problem pertains to the organic materials in solution, because they have the reputation of being less efficient than "evaporable" materials.

Another problem is linked to the cost of inkjet deposition equipment. This is primarily due to the cost of developing specific printing heads with a relatively short life cycle, along with the price of all the chemical products needed for the preparation of the solutions and the processes of cleaning of the system.

2.6.4. The technique of Roll-to-Roll deposition

Roll-to-Roll deposition is a very promising production method, as it could prove to be the least expensive way to make OLED panels, which are of particular interest for applications in lighting. This technique encompasses a set of techniques, using rollers to deposit organic materials in solution (Figure 2.23). These procedures use a flexible substrate roll which may be very long indeed.

Figure 2.23. Left: a Roll-to-Roll printer. Right: a flexible OLED made by the Roll-to-Roll technique

2.6.5. What is the best deposition method?

The above list of OLED deposition/manufacture techniques is, of course, non-exhaustive. There are numerous other techniques that can also be used to create organic optoelectronic devices. The choice of technique depends on two considerations:

– the physical and chemical properties of the organic semiconductor used;

– the final device and the intended application (lighting, display, large surface, etc.).

The current expansion of OLED use in the fields of lighting and display should stimulate this domain of study.

2.7. Characterization of an OLED's electroluminescence

Measuring the optical properties of an OLED is a very important task, in order to obtain a figure of merit and sometimes evaluate the device's viability. An OLED has a 2D emission surface; it is very thin, can be deposited on any type of substrate, and exhibits a large emission angle, which can even be up to 360° (if we use a transparent substrate). These various advantages mean that it is a difficult and delicate task to choose the method of optical characterization. Particular attention must be paid to the level of uncertainty and inconsistency of the results reported in the scientific literature hitherto published on OLEDs. This necessitates that we first define, and then establish methods and normalized values to better define the electroluminescence of an OLED.

Two types of physical values can be collected, depending on the intended application for the light source: the physical values relating to the luminous radiation, defined on the basis of energy considerations and describing the energetic photometry (radiometry), and those defined taking account of the stimulation of the human eye, describing the visual photometry. It is this photometric measurement which is most commonly employed in the lighting and display industries (see Chapter 1).

2.8. Current-voltage-luminance (J-V-L) characterization of an OLED heterostructure

Characterizing an OLED consists mainly of obtaining the family of curves showing current density as a function of the applied voltage and the luminance as a function of the voltage – examples are shown in Figure 2.24. At each voltage step, we measure the current injected and the flux emitted by the OLED. The current is converted to current density to take account of the active surface area of the OLED. The flux is converted to luminance (cd/m²) in line with the explanations given in Chapter 1.

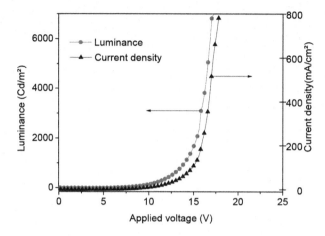

Figure 2.24. *J-V-L characteristics of an OLED*

A log-log representation of the current density is shown in Figure 2.25 [RAM 06]. This representation enables us to identify the different regions on the basis of the characteristic current density as a function of the applied voltage.

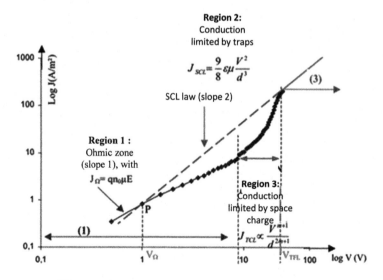

Figure 2.25. *The log(J) curves as a function of logV in the cases of shallow and deep traps [RAM 06]*

In an I-V curve, we distinguish a region below the voltage, an operational region and a saturation region.

First of all, the current is essentially Ohmic, and there is no light emission. In this zone, the current density varies in a linear fashion with the applied voltage, and it obeys an Ohmic law, of the form:

$$J_\Omega = q\,n_0\,\mu\frac{V}{d}$$

[2.11]

This intrinsic, so-called Ohmic conduction is caused thermally by the residual free charges (which do not, *a priori*, include the trap levels), whose concentration remains greater than that of the injected charges (V very low: no injection).

In the second zone, the voltage applied is high enough for the space charge density to be sufficiently great throughout the device to generate light. However, a great many of these charges, as a matter of priority, fill the trap centers (non-radiative centers). In this region, the current is proportional to the applied voltage (8-12 V), and the luminance increases considerably with rising voltage. The variation in current density is described by the so-called trapped-charge limited (TCL) law [STÖ 99], in the form:

$$J_{TCL} \propto \frac{V^{m+1}}{d^{2m+1}}$$

[2.12]

The transition between the above two regions occurs at the threshold voltage.

When all the traps are filled ($V = V_{TFL}$), the saturation level is reached, which causes a transition from the TCL regime, with trapped charge carriers, exponentially to a space-charge limited current (SCLC) regime, where the carriers are not trapped [RAM 06]. The current density then obeys an SCL law which corresponds to the SCLC in a material without traps. In this case, the current density is given by the relation:

$$J_{SCL} = \frac{9}{8}\varepsilon\mu\frac{V^2}{d^3}$$

[2.13]

OLED performances are generally characterized by:

– the slope of the current curve as a function of the voltage (in the high-voltage domain);

– the working voltage – the voltage beyond which the luminous intensity is greater than a certain level of luminance (usually around 1 cd/m²);

– the maximum luminance emitted by the OLED at a given level of voltage;

– the maximum quantum and luminous yields;

– the lifetime of the OLEDs, which is the operating time t for which the luminance can be maintained at a value greater than half its original value (its value at $t = 0$), set in principle at 100 cd/m².

2.9. Conclusion

In this chapter, we have recapped the working principle behind OLEDs (the equivalents of LEDs, but using organic materials). Multiple parameters need to be optimized in order to increase an OLED's overall yield. The R&D activities conducted over the past few decades have enabled us to obtain efficient OLEDs, used in the domains of lighting and display. Since the earliest work, in 1987, by Tang and Van Slyke on an OLED comprising two organic layers (copper phthalocyanine: CuPC, and triquilonine aluminum: Alq3), placed between an ITO anode and a cathode made of a silver–magnesium alloy, numerous successes have been achieved. Yet there are still many challenges to overcome to further improve the performances of OLEDs.

The energy levels of the materials used in OLED stacks play an important role in the efficiency of the OLED. The materials are chosen so that the difference in level between the injection- and transport layers is minimized, and so that only the barrier at the recombination interface remains. With such structures, the yields can be up to 3.5% external quantum yield, and almost 25% internal quantum yield, which is the theoretical limit with fluorescent materials. Such OLEDs provide typical performances of 1000 cd/m² and current densities of 35 A/cm² in continuous operation. This offers a glimpse of prospects which can be interesting for applications, and above all, for the quest for organic laser diodes, mirroring the process of turning an OLED into an OLD (Organic Laser Diode).

3

Organic Lasers

Organic lasers, and particularly dye lasers, represent only one portion of the different laser technologies available. The reason why, today, there is a renewed interest in organic lasers is that organic semiconductors are very flexible and cheap to produce. In addition, thanks to the availability of an almost limitless number of organic molecules, we can obtain emissions across the entirety of the visible spectrum, and even a part of the near infrared and deep blue.

Organic lasers date back almost as far as lasers in general: while the first laser was made in 1960 [MAI 60], the first dye laser – i.e. the first one to use a gain medium made of organic material – was demonstrated in 1966. That year, Sorokin and Lankard at IBM were the first to obtain stimulated emission from an organic compound – specifically, aluminum phthalocyanine chloride. Strictly speaking, in view of its central metal atom directly connected to organic molecules in a ring, this dye is made up of organometallic molecules. Given that dye lasers and their scientific and technical evolution were the first steps in the field of organic lasers, it seems pertinent to re-examine this history of dye lasers, and cross-reference it with the history of OLEDs to gain a better perspective of the possibility of creating an organic laser using electrical pumping.

3.1. Principle behind lasers

3.1.1. *Transition mechanisms*

The critical point in the making of a laser, whatever the pumping mechanism, is amplification by stimulated emission ("laser" actually stands for *light amplification by stimulated emission radiation*). This effect was postulated by A. Einstein in 1917, and exploited by T. Maiman in 1960 to create the first ruby laser.

To amplify the light, it is crucial to invert the population – that is to say, the population of emissive excited states must be greater than the population of the ground states.

Consider a two-level gain medium (illustrated in Figure 3.1) with the population of atoms (or molecules) in the ground state being N_1, and the population in the excited state N_2. Generally, three phenomena may occur in such a system: absorption, spontaneous emission and stimulated emission.

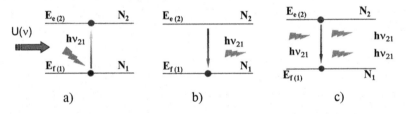

a)   b)   c)

Figure 3.1. *Transition mechanisms in a two-level system: a) absorption, b) spontaneous emission and c) stimulated emission*

In the case of spontaneous emission (see Figure 3.1(a)), an electron can move from an energy level E_1 to energy level E_2 by absorbing energy from incident radiation at a frequency ν centered around ν_0.

The number of transitions from 1 to 2 per second, per unit frequency and in the unit volume is given by:

$$\frac{dN_1}{dt} = -B_{12}N_1U(\nu) \qquad [3.1]$$

N_1: population of level 1

B_{12}: probability of absorption per second per frequency (Einstein coefficient)

$U(n)$: density of incident energy [Js/m^3]

The de-excitation of this system takes place firstly by a spontaneous emission (Figure 3.1(b)). The rate of spontaneous emission is N_2A_{21}, such that:

$$\frac{dN_2}{dt} = -A_{21}N_2$$ [3.2]

A_{21}: Einstein coefficient

$t_2 = 1/A_{21}$ is the lifetime of the excited state (10^{-9}-10^{-7}s). It corresponds to the time after which the population density drops by $1/e$.

Under the influence of the incident electromagnetic (EM) radiation, an electron can also move from state 2 to state 1 by ceding a photon of the same frequency, direction and polarization as the incident radiation (Figure 3.1(c)). This is stimulated emission. The rate of stimulated emission is $\rho N_2 B_{21}$, where ρ is the energy density in the medium. It is given by:

$$\frac{dN_2}{dt} = -B_{21}N_2\rho(v)$$ [3.3]

At thermodynamic equilibrium, we have:

$$\rho N_1 B_{12} = N_2 A_{21} + \rho N_2 B_{21}$$ [3.4]

From this, we deduce:

$$\rho = \frac{A_{21}/B_{21}}{\dfrac{B_{12}N_1}{B_{21}N_2} - 1}$$ [3.5]

and Boltzmann statistics gives us:

$$\frac{N_2}{N_1} = exp\left(-\frac{E_2 - E_1}{k_B T} \right)$$ [3.6]

According to Planck's law, the distribution of the spectral energy luminance of the black-body thermal radiation at thermal equilibrium is:

$$I = \frac{2h\upsilon^3}{c_m^2} \frac{1}{exp\left(\dfrac{h\upsilon}{k_B T} \right) - 1}$$ [3.7]

v_m: velocity of light in the material

$h\upsilon$: energy of the radiated photon

However, $\rho = 4\pi I/c_m$ and $E_2 - E_1 = h\upsilon$.

By identification with equation [3.2], adopting the hypothesis of homogeneous emission in the solid angle 4π, we obtain:

$$B_{12} = B_{21}$$ [3.8]

$$\frac{A_{21}}{B_{21}} = \frac{8\pi h\upsilon^3}{c_m^3}$$ [3.9]

This indicates that the probability of absorption is equal to the probability of stimulated emission.

In a homogeneous medium, the optical intensity at position z can be written as:

$$\frac{dI}{dz} = \frac{1}{4\pi}\left(N_2 - N_1 \right) B_{21} \rho h\upsilon$$ [3.10]

Let us posit that $\alpha = -\left(N_2 - N_1\right)B_{21}h\upsilon / c_m$. This being the case, we can rewrite the above equation as:

$$\frac{dI}{dz} = - \alpha I \qquad\qquad [3.11]$$

The solution to this equation gives us the definition of the absorption coefficient (α), such that:

$$I = I_0 e^{-\alpha z} \qquad\qquad [3.12]$$

I_0: intensity at the origin ($z = 0$)

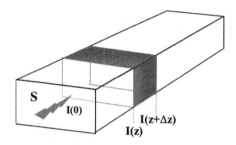

Figure 3.2. *Propagation of a wave in a medium*

At ambient temperature and in the absence of external excitation, we have $N_2 < N_1$ and $\alpha > 0$. Consequently, the intensity decreases as we advance in direction z, meaning that the medium is absorbent, and α is the absorption coefficient.

To produce amplification, we need to act on the medium so that $N_2 > N_1$ – that is to say that the medium becomes "active" when *population inversion* $\Delta N = N_2 - N_1 > 0$ is caused by an external excitation. In this case, $\alpha < 0$ and $I(z) > I_0$. The latter is the optical gain, and the symbol α is replaced with g.

Considering the definition of the effective stimulated emission cross section $\sigma_{sti} = B_{21}h\upsilon / c_m$, the optical gain is the product of σ_{st} by ΔN:

$$g = \Delta N \sigma_{sti} \qquad\qquad [3.13]$$

However, it must be noted that it is impossible to affect a population inversion in the case of a two-level system, and therefore we cannot amplify radiation with this type of system.

The solution is to use a multi-level system – particularly three-level or four-level systems (Figure 3.3).

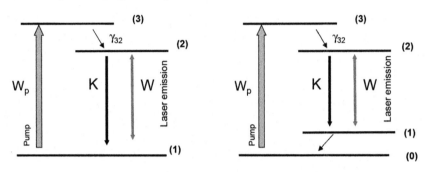

Figure 3.3. *Diagram of the principle behind a 3- and 4-level system (left and right, respectively). W_p, K, W and γ_{32} are the transition probabilities*

Note that the four-level system is more efficient than a three-level one. The four-level system is capable of continuous operation, and does not require a transparency threshold.

For a four-level system (represented by E_0, E_1, E_2 and E_4 in Figure 3.3.(b)), $\Delta N > 0$ is obtained when the system is pumped, because E_1 and E_2 are initially (before the application of an excitation) unoccupied states. Hence, it is an ideal system from which to make a laser.

The organic emitter – of which the example used mainly for this book will be Alq3:DCM – can be considered to be a four-level system [BRÜ 11], even if the HOMO and LUMO levels of the two types of molecules are quasi-contiguous bands. ΔN in the Alq3:DCM, therefore, is the singlet population density.

As explained in Chapter 2, organic materials have the peculiarity of exhibiting multiple vibrational levels in the ground state S_0 and the excited state S_1. The energy levels of an organic material are illustrated in

Figure 3.4. This figure shows the ground state S_0 and the first excited state S_1 with their respective vibrational levels. We can see the similarity with conventional 4-level lasers. Consequently, the energy levels of organic materials mean that they can more accurately be described by a four-level model [SCH 98].

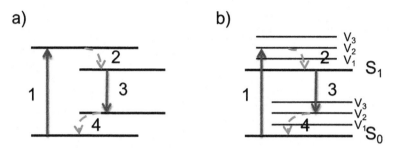

Figure 3.4. *Energy diagram of a conventional four-level system in a) and for an organic material in b), with the optical absorption at 1, radiative emission at 3 and non-radiative vibrational relaxation between V_2 and V_0 in S_1 and between V_2 and v_0 in S_0*

In addition, organic materials have a broad spectral absorption range, and the emission is red-shifted in relation to the absorption band. Hence, reabsorption losses, at the emission wavelength, are considerably reduced. Also, organic materials are good candidates for use as active materials. Depending on the organic compound, it is possible to obtain an emission covering the whole of the visible spectrum.

The above models are ideal systems with finite energy levels, and the emitted photon is at the single frequency of υ, resulting in a single, infinitely narrow line on the spectrum. However, in the real world, systems do not have spectral lines like this – the lines may be broadened or shifted on the spectrum due to numerous physical mechanisms, such as natural broadening, the Doppler effect, intermolecular interaction, etc. In view of these effects, the gain of a true material varies as a function of the electromagnetic wave frequency (or the equivalent in wavelength), and is written as $g_{mat}(\upsilon)$.

The laser is triggered when the gain produced is capable of compensating all optical losses in the laser cavity. The threshold of a laser, therefore, is essentially defined by the *threshold population density* or *threshold gain*.

3.1.2. *The laser cavity*

In a conventional Fabry–Pérot cavity [KOS 05], the gain medium is placed between two mirrors, and the photons travel back and forth, traversing the gain medium multiple times (Figure 3.5). Thus, the effective length of the gain medium is a multiple of the real length of the cavity, and the total amplification is increased in comparison to a system where the photons only pass through the gain medium once. At the same time, the cavity is selective for photons of different frequencies, and the spatial distribution of gain and losses is not uniform within the cavity. For certain frequencies, all of the photons are confined in the cavity, producing steady-state waves known as the "modes" of the cavity. We distinguish between longitudinal and transverse modes [AKS 14]. Longitudinal modes are distinguished solely by their frequency, and transverse modes by their spatial distribution. In a plane–plane Fabry–Pérot cavity, the longitudinal modes are characterized by:

$$v_i = \frac{ic}{2L_{opt}} \quad \text{where } i \in \mathbb{N} \qquad [3.14]$$

$L_{opt} = \sum_i n_i l_i$ is the optical length of the cavity and c the celerity of light

The frequencies v_i are the frequencies of the peaks on the spectrum.

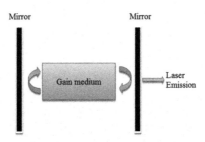

Figure 3.5. *Diagram of a Fabry–Pérot laser cavity*

Consider a Fabry–Pérot laser cavity, containing an active zone and a passive zone (Figure 3.6). The electric field of the steady-state wave on the axis of propagation is a sine function. Supposing we have a transverse distribution expressed by $\hat{E}_T(x,y)$, the electric field in the cavity is given by:

$$E_n(x,y,z) = \hat{E}_T(x,y)\sqrt{2}\cos\left(\frac{2\pi n}{\lambda}z\right)$$

[3.15]

$\hat{E}_T$: profile of the normalized transverse electric field

n: refractive index

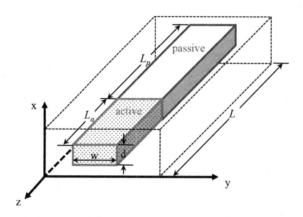

Figure 3.6. *Laser cavity with an active zone and a passive zone in the direction of laser propagation (z axis)*

The modal gain of the mode described by this electric field distribution is:

$$\langle g_{mod}\rangle = \frac{\iiint_{xyz} g_{mat}(x,y,z)|\hat{E}_T(x,y)|^2 2\cos^2\left(\frac{2\pi n}{\lambda}z\right)dxdydz}{L\iint_{xy}|\hat{E}_T(x,y)|^2 dxdy}$$

[3.16]

Equation [3.16] illustrates that the gain of the mode is given by the gain of the material g_{mat}, weighted by the squared distribution of the electric field (intensity). Supposing that the gain and absorption are homogeneous in the different zones making up the cavity, the modal gain is thus written:

$$\langle g_{mod} \rangle = \Gamma_{mod}^{xy} \Gamma_{mod}^{z} g_{mat} \qquad [3.17]$$

where Γ_{mod}^{z} is the longitudinal confinement factor and Γ_{mod}^{xy} is the transverse confinement factor. These two factors can be computed by:

$$\Gamma_{mod}^{xy} = \frac{\int_{-w/2}^{w/2} \int_{-d/2}^{d/2} \left| \hat{E}_T (x,y) \right|^2 dxdy}{\iint_{-\infty}^{\infty} \left| \hat{E}_T (x,y) \right|^2 dxdy} \qquad [3.18]$$

$$\Gamma_{mod}^{z} = \frac{1}{L} \int_{-L_a/2}^{L_a/2} 2cos^2 \left(\frac{2\pi n}{\lambda} z \right) dz = \frac{L_a}{L} \left[1 + sinc \left(\frac{2nL_a}{\lambda} \right) \right] \qquad [3.19]$$

$$sinc(x) = sin(\pi x) / \pi x \qquad [3.20]$$

For a plane–plane Fabry–Pérot cavity of large dimensions (the width w and height d are large), the transverse confinement Γ_{mod}^{xy} is close to 1. The longitudinal confinement is mathematically the ratio of the thickness of the active layer (L_a) to the total thickness of the cavity (L) with weighting by the spatial distribution of the electric field.

In the case of organic lasers, the position of the active layer and passive layers in relation to the spatial distribution of the steady-state wave must be optimized in order to maximize the optical confinement, and thus maximize the modal gain. This phenomenon may play an important role, particularly in the case of an active medium of an OLED, which is a very thin layer < 100 nm, less than the emission wavelength. The position of the layer of the active medium must therefore be on a peak of the steady-state wave intensity, and the absorbent layers in the troughs, in order to maximize gain and minimize losses.

As a function of the resonant cavity, the laser oscillation is limited to resonant modes. Only photons whose direction of propagation is aligned with the axis of the cavity amplify the emission.

It is difficult to determine when the laser threshold is reached. In general, the following characteristics are associated with the presence of laser emission [SAM 09]:

– the presence of a pump energy threshold as a function of the emission energy;

– a coherent emission beam;

– spectral narrowing;

– the presence of a resonant cavity.

Before discussing the different types of laser resonators, it is important to note that the small dimensions of the organic layers generally impose the use of micro-resonators.

Optical microcavities (or micro-resonators) are resonators with at least a dimension of the order of a micrometer, which is comparable with the wavelength of the light. Confinement of light in a small volume leads to the modification of the emission properties of the emitters in the microcavity. We can expect an enhancement or inhibition of spontaneous emission depending on its polarization, its wavelength or its spatial orientation.

Various kinds of microcavities have been developed, including the Fabry–Pérot VCSEL (Vertical-Cavity Surface-Emitting Laser), the DFB (Distributed Feedback Bragg Mirror), the 2D photonic crystal microcavity, the microdisk, etc.

One of the initial motivations for making microcavities was to grow the spectral interval so that the laser can operate in monomode. Laser thresholds are greatly reduced due to the modification of the spontaneous emission by the Purcell effect [STR 11, PEL 02].

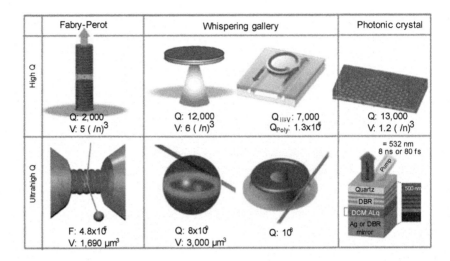

Figure 3.7. *Left: examples of microcavities with a high quality factor and small modal volume. Top row: micropill, microdisk, 2D photonic crystal; bottom row: Bulk Fabry–Pérot cavity, microsphere and microtore [SEI 07]. Bottom right figure: studies of VCSEL Alq3:DCM microcavities conducted by G.M. Akselrod [AKS 14]*

Indeed, the modification of the spontaneous emission rate can be described by the Purcell factor, which is the ratio between the lifetime of spontaneous emission outside of the cavity (τ_{sp}^0) and that in the cavity (τ_{sp}^c) [STR 11]:

$$F_P = \frac{\tau_{sp}^0}{\tau_{sp}^c} \qquad [3.21]$$

At the resonance wavelength of the microcavity, the emission efficiency of an emitter is enhanced by a Purcell factor, given by:

$$F_P = \frac{3}{4\pi^2}\left(\frac{\lambda}{n}\right)^3\left(\frac{Q}{V}\right) \qquad [3.22]$$

λ/n: wavelength in the medium

Q: quality factor

V: modal volume

In the case of conventional lasers, whose dimensions are far greater than the wavelength, multiple modes interact with the active medium. Consequently, the spontaneous-emission coupling factor in laser mode (β), i.e. the rate of coupled spontaneous emission in laser mode in relation to the global emission, is small – typically of the order of 10^{-6} [STR 11].

In a VCSEL microcavity, the spontaneous-emission coupling factor in laser mode (β) is approximately calculated by [KAK 02]:

$$\beta = \frac{F_p}{1 + F_p} \qquad [3.23]$$

A large Purcell factor, therefore, results in a higher spontaneous-emission coupling rate. It has been shown that the threshold laser, often defined by the turning point in the intensity curve at the output as a function of the pumping intensity (input–output curve) decreases as a function of the factor β [BAB 03].

It has been suggested that a "thresholdless laser" could be created if $\beta = 1$. This "prediction" is called into question by physicists in quantum optics [KUM 12, NIN 13], because the input–output curve is not sufficient to define the lasing threshold [STR 11]. Indeed, the triggering of the laser is experimentally characterized by multiple "signatures": (1) the emission spectrum exhibits narrow peaks; (2) the output light is a beam; (3) a threshold is observed for the output intensity and for the narrowing of the emission spectrum [SAM 09]. In other words, the lasing threshold must be able to distinguish the spontaneous emission regime and the stimulated emission regime. Conventional lasers exhibit a clear boundary between these two regimes and microcavities with a large factor β tend rather to exhibit a wide transition zone, meaning that the lasing threshold is difficult to define.

From the point of view of reducing optical losses in the microcavity, the quality factor Q plays an essential role in reducing the laser threshold. For the VSCEL organic laser, [AKS 14] reported the study of two similar microcavities having a different quality factor: DBR-Ag ($Q = 300$) and DBR-DBR ($Q = 3000$) with the same thickness of the active medium. With excitation by a nanosecond laser, the lasing thresholds are, respectively, 80 $\mu J/cm^2$ and 4 $\mu J/cm^2$, and with excitation by a femtosecond laser, they are 4 $\mu J/cm^2$ and 0.4 $\mu J/cm^2$. Thus, we observe that the lasing threshold is smaller if Q is higher.

3.1.3. *Pumping*

3.1.3.1. *Optical pumping*

Optical pumping was first developed, in terms of theory, in 1950 by A. Kastler, and actually realized for experimentation in 1960 by T.H. Maiman. Initially, a pulse discharge lamp was used to obtain an intense flash of light in a broad spectrum (white and UV light). The process is repetitive with a frequency of a few Hz to a few tens of Hz.

However, this mode of pumping offers a low yield, given the spatial dispersion of the light and the presence of unusable frequencies.

To remedy this problem, the pumping of a gain medium by another laser yields better results.

3.1.3.2. *Electrical pumping*

We can distinguish two types of electrical pump: electric discharge pumping of a fluid medium and pumping by current injection, notably in a solid medium.

In the first case, electrons that are greatly accelerated by an electrical discharge cede a part of their kinetic energy to the atoms or molecules of a gas at reduced pressure. The energy transfer can induce the excitation of certain energy levels (electronic in the case of a monatomic gas or rotation/vibration in the case of a molecular gas).

3.2. Laser effect in organic materials

Many groups have launched research activities into organic lasers, and particularly organic diode lasers (electrically pumped). As previously indicated, the first organic laser in solution, also called a dye laser, was demonstrated in 1966 by P. Sorokin [SOR 66]. In 1992, D. Moses demonstrated the laser effect, under optical pumping, in a solution based on a polymer: poly(2-methoxy-5-(2'-ethylhexyloxy)-1.4-phenylene vinylene) (MEH-PPV) [MOS 92]. It was only in 1996 that the laser effect in an organic layer was demonstrated by F. Hide [HID 96] (again with optical pumping).

Since then, organic lasers have been demonstrated for the whole of the visible spectrum, for various classes of materials and for different geometric forms of the resonator, but always with pumping by optical means. The groups lead by Chen at Caltech (USA) [CHE 07] and Kitamura at the University of Tokyo (Japan) [KIT 05] are among the most active. In China, the opto-electronics laboratory at the University of Changchun [DON 08] is also carrying out research in this domain. At European level, scientific research on the organic diode laser is also highly active. In the United Kingdom, I. Samuel's group at the University of St. Andrews worked on DFB organic lasers [VAS 05, VAS 06]. In Germany, the group directed by K. Leo at Dresden University [KOS 05] plays a leading role in work on vertical microcavities. Research at the University of Denmark has also included study on the laser effect in organic materials [CHR 08]. In France, the quest for an organic laser is one of the main research goals of the PON team (led by A. Boudrioua) at the laser physics laboratory at Paris 13 University [CHA 11].

The laser effect has been shown with different types of organic materials and in various resonant configurations, and this work will be reviewed in the coming sections.

3.2.1. *Optical gain in organic semiconductors*

A precondition for the laser effect is the presence of stimulated emission, which is quantified by the stimulated emission effective cross-section $\sigma_{SE}(\lambda)$, which, itself, is dependent on the emission wavelength. It is given by:

$$\sigma_{SE}(\lambda) = \frac{\lambda^4 f(\lambda)}{8\pi n^2 c \tau_{rad}}$$

[3.24]

where $f(\lambda)$ is the normalized photoluminescence distribution, n is the refractive index of the material, "c" is the celerity of light and τ_{rad} is the radiative lifetime.

It is preferable to use a laser material with a high gain; this is expressed by a large stimulated emission effective cross-section. Such is the case with organic materials whose order of magnitude of $\sigma_{SE}(\lambda)$ is 10^{-16} cm² [SCH 97].

Organic materials exhibit strong stimulated emission cross-section for the radiative transition from the singlet state S_1 to the ground state S_0.

Generally, the luminous intensity propagating in a gain medium can be written as:

$$I = I_0 e^{(g(\lambda)-\alpha)z}$$

[3.25]

where g is the gain of the material, I_o is the initial intensity, α is the loss coefficient and z is the length of the gain medium. The gain is given by:

$$g = \sigma_{SE}(\lambda)N$$

[3.26]

where $\sigma_{SE}(\lambda)$ is the effective stimulated emission cross-section and N is the density of excited states.

These factors determine the overall gain of an optically pumped organic material. One of the important criteria for a gain medium is spectral non-overlap between stimulated emission and the absorption of the organic material. Another phenomenon giving rise to absorption is the intersystem crossing and the population of the triplet states. The absorption of the triplet states is particularly significant in continuous pumping due to their sufficiently long lifetime.

One of the methods to measure the gain in the organic materials consists of analyzing amplified spontaneous emission (ASE). ASE is the interaction of the photons emitted by the singlet states excited along the optical path in the gain material. To measure the gain, a thin organic layer deposited on a substrate with a low index is pumped by pulsed light over a small area near to the edge. The measurement is made per slice of the sample. The light thus emitted is guided along the sample. Spontaneous emission is guided and then amplified until it reaches the edge of the sample. In these conditions, the luminous intensity of the material is amplified, leading to a narrowed spectral emission beyond a certain pump intensity [CAL 10]. The output intensity, $I(\lambda)$, of the ASE is given by the relation:

$$I(\lambda) = \frac{A(\lambda)I_p}{g(\lambda)}[\exp(g(\lambda)l - 1]$$

[3.27]

where $A(\lambda)$ is a constant linked to the emission section, I_p is the pump intensity and "l" is the length of the pump zone.

Therefore, by controlling the emission intensity as a function of the length of the pump zone, we can calculate the gain $g(\lambda)$. This method was initially applied to inorganic semiconductors [SHA 71], and it was then adapted to organic materials [LU 04, JOR 06].

In addition, by moving the slit as a function of the waveguide edge, it is possible to measure the propagation losses in the waveguide. Note that the losses in organic materials are typically of the order of 3-50 cm^{-1}.

To obtain the laser effect, an organic material therefore needs to have low reabsorption, a large stimulated emission effective cross-section and low ISC. The advantage to having a high laser gain is that it enables us to reduce the size of the laser.

3.2.2. *Optical resonators*

As previously mentioned, a laser needs two elements to function. The first is a gain medium and the second a resonant structure to amplify the signal and thus obtain a laser emission. A wide variety of experiments reported in the literature show different types of laser resonators, based on organic materials with optical pumping. The different types of resonators can be divided into sub-categories, which are outlined below:

– vertical cavities [KOZ 98];

– micro-rings or micro-discs [ADA 08];

– DBR planar resonators [VAS 06];

– 1D DFB photonic crystals [HEL 04];

– 2D DFB photonic crystals [CHR 08, DON 08, BAU 07, HAR 05, HEL 04, HEL 03];

– photonic crystal microcavities [KIT 05, MUR 10].

3.3. Theoretical model of an organic semiconductor laser

In this section, we focus on the emission properties of organic materials in a microcavity. The microcavity alters the emission properties of the medium by a factor β. As previously mentioned, this parameter characterizes the proportion of spontaneous emission coupled with a resonant mode of the cavity. In a laser, it has been shown that, if the coupling ratio of the spontaneous emission in the laser modes of a cavity is sufficiently high, the properties of the laser effect are changed – notably the threshold. An additional view is that it is the result of the improvement of the resonance of the EM field by multiple reflections in the cavity. The improvement of emission is greater when the length of a return journey of the light is short in relation to the lifetime of a dipole. The cavity effectively increases the coupling of the dipole to the field.

The characteristics of a laser are generally described by the so-called "laser-rate-equations". However, these equations are valid only on condition that the intensity of the pump or of the light absorbed is directly proportional to the emission intensity, which is so only when there are a large number of molecules in the modal volume of the laser. This hypothesis or condition is reasonable for a conventional laser with a significant modal volume, which corresponds to a large number of molecules that can be excited. Additionally, in the case of a photonic crystal microlaser cavity, due to a much smaller modal volume than with a conventional laser, it contains very few molecules per laser mode, and thus its maximum available gain per mode is very low. To account for the small size of the laser, two terms need to be included in the equations for the laser:

– the Purcell factor (F) to take account of the gain of the spontaneous emission at the wavelength of the laser mode;

– the number of absorbent molecules per laser mode.

In addition, semi-classic laser equations may be capable of describing the behavior of organic lasers. These equations take account of the improvement of spontaneous emission caused by the microcavity. Supposing we have an ideal four-level laser material and taking account of the non-radiative processes of depopulation. The equations are:

$$m_{tot} = m_0 + m \qquad\qquad [3.28]$$

$$\frac{dm}{dt} = p_e(m_0 - m) - Km \qquad\qquad [3.29]$$

$$\frac{dn}{dt} = Km - (1-\beta)A_f n + \beta(1+q)A_f n - \Gamma n \qquad\qquad [3.30]$$

$$\frac{dq}{dt} = \beta A_f(1+q)n - \gamma q \qquad\qquad [3.31]$$

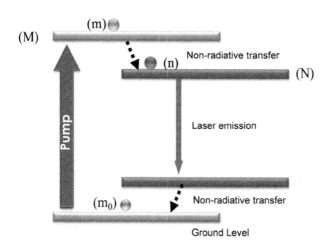

Figure 3.8. *Diagram of the energy diagram of a conventional four-level system*

Equations [3.28] to [3.31] represent the evolution of the population of the levels M and N and the number of photons (q) in the laser mode with n and m representing the number of molecules of the excited state at levels N and M, respectively (Figure 3.8). A_f represents the spontaneous emission rate modified in a microcavity. m_{tot} is the total number of molecules and m_0 is the number of molecules in the ground state. The coupling rate of spontaneous emission in the laser mode is called β, γ describes the rate of loss of the photons in laser emission mode and Γ represents the rate of non-radiative recombination of the excited states (including losses by bimolecular intersection). K represents a set of parameters such as the vibrational relaxation rate, or the Förster energy transfer rate.

P_e is the pumping rate of the system; it is the rate of excited molecules generated in laser mode per unit time, and is expressed by:

$$P_e = \frac{P_{in}}{S_M h v} \sigma_{abs}$$

[3.32]

where P_{in} is the total light power of the pump, S_M is the surface of the pump spot.

The model may be made more accurate if we also take account of the triplet excitons and the losses linked to the molecular interactions. The following equations determine the density of singlet excitons, triplet excitons and photons in the presence of stimulated emission:

$$\frac{dm}{dt} = P_e(m_0 - m) - Km$$

[3.33]

$$\frac{dS}{dt} = Km - (1-\beta)A_f S + \beta(1+q)A_f S - \kappa_{SS}S^2 - \kappa_{STA}ST + \frac{1}{4}\kappa_{TTA}T^2 - \kappa_{ISC}S$$

[3.34]

$$\frac{dq}{dt} = \beta A_f(1+q)S - c(\alpha_{cav} + \sigma_{TT}N_T)q$$

[3.35]

$$\frac{dT}{dt} = \kappa_{ISC}S - \frac{5}{4}\kappa_{TTA}T^2$$

[3.36]

Here, S and T are, respectively, the densities of singlets and triplets, q is the density of the photons in laser mode and α_{cav} represents the losses of the cavity.

For example, Figure 3.9 shows the evolution of densities of excitons as a function of time. We can see that the density of polarons in the structure becomes saturated 30 ns after the current is applied. At a low current density, the population density of singlet states is mainly determined by the recombination rate and the rate of de-excitation of the singlet states κ_S.

When the density of triplet states increases, the annihilation rate limits the population of singlet states.

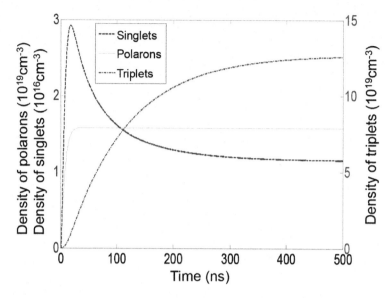

Figure 3.9. *Simulations of the population densities of the singlet states, triplet states and polarons*

3.4. Optically pumped organic lasers

3.4.1. *The organic gain medium*

As we can see from the previous sections, organic semiconductors are described by their HOMO and LUMO levels. On its own, consideration of these two levels does not justify a possible laser effect. Indeed, a two-level system renders population inversion impossible. In order to justify a possible laser amplification in organic semiconductors, it is necessary to consider levels higher than the LUMO, and levels lower than the HOMO (Figure 3.10). As Figure 3.10 shows, the HOMO level has lower energy sublevels, rendering possible the depopulation of the excited states in level E_1. Therefore, there will be more excited states in level E_2 than level E_1, fulfilling the condition of population inversion, which is indispensable to encourage stimulated emission and thus obtain a laser effect.

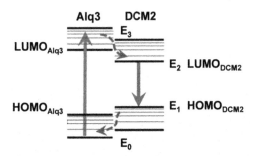

Figure 3.10. *Diagram of the energy levels of a 4-level laser depending on an Alq3-DCM$_2$ guest-host system*

As previously indicated, as the gain $g(\lambda)$ of organic materials is proportional to the population inversion and to their effective cross-section: the larger the effective cross-sections and the population of excited states, the greater the gain will be. The effective cross-section of organic materials is around 10^{-16} cm² [SCH 97]. Taking account of the loss coefficient α, the gain of the organic material $g(\lambda)$ and the length of the cavity z, the intensity of the laser beam varies exponentially with the distance z, in accordance with formula [3.25]. The losses in solid-state organic materials are of the order of 1 cm^{-1} [BER 97].

We generally see big differences between the measured and calculated values. Let us take the example of DCM (4-dicyanmethylene-2-methyl-6-(p-dimethylaminostyryl)-4H-pyran). S. Riechel *et al.* [RIE 01] used relation [3.27] to calculate a gain value $g_{DCM} \approx 700$ cm^{-1}, considering the number of excited molecules $N = 5 \times 10^{19}$ and a spontaneous emission effective cross-section $\sigma_{SE} = 1.3 \times 10^{-17}$ cm². On the other hand, S. Toffanin *et al.* [CAP 10] experimentally measured a DCM gain of $g_{DCM} = 77$ cm^{-1}. In spite of these disparities between theory and practice, DCM is considered to be a good gain medium, which is very often used in optically pumped organic lasers, notably for its important gain, but also for its excellent Förster transfer with its host matrix (see section 1.1.8.1).

Let us now look at different types of optical cavity having been used in making optically pumped organic lasers.

3.4.2. *Different types of laser cavity*

In this section, we present a brief state of the art on organic lasers using optical pumping, which will open the doors for us to then examine the issue of making an electrically pumped laser diode.

3.4.2.1. *Fabry–Pérot laser*

Historically, Fabry–Pérot cavities were the first to have been demonstrated. The gain material (organic material) is placed between two mirrors: one highly reflective, and the other semi-transparent. For a monochromatic plane wave whose intensity is I_0 with an incidence i, the wave transmitted depends on the difference in optical path, which corresponds to a two-way journey in the cavity. Thus, we can define the Airy function, such that:

$$\frac{I}{I_{max}} = \frac{1}{1 + m \sin^2\left(\dfrac{\varphi}{2}\right)} \quad where \ m = \frac{4R}{(1-R)^2} \qquad [3.37]$$

where I_{max} is the maximum transmitted intensity, R is the reflective coefficient of each of the two faces and φ is the phase mismatch. Remember that m increases as the reflective coefficient of the mirrors R grows.

In this configuration, lasers with a minimum threshold reported under optical pumping are in the range of 30-400 kW/cm² with quality factors of the order of $Q = 420\text{-}4500$ and a line width of up to 0.05 nm (400 kW/cm² and $Q = 4500$ [KOS 05]).

3.4.2.2. *DFBs (distributed feedback lasers)*

Distributed feedback lasers (DFBs) emerged in the 1970s (Figure 3.11). A complete description of the mechanism involved was given by Kogelnik and Shank [KÖG 71]. In this case, the use of a Bragg lattice in an active medium produces a single-mode laser emission.

A typical structure of the DFB organic laser consists of depositing a layer of organic material on a periodic structure. The light propagating in the waveguide undergoes backscattering within the lattice. This backscattering, on each step of the lattice, causes constructive interference. The condition of phase tuning only occurs for certain wavelengths corresponding to Bragg's condition of the lattice, which stipulates that the wavelengths in the lasing medium are linked to the period of the lattice by the following relation [KÖG 72]:

$$m\lambda = 2n_{eff}\Lambda \qquad\qquad [3.38]$$

where λ is the laser wavelength, Λ is the step of the lattice, m is an integer representing the order of diffraction, and n_{eff} is the effective refractive index of the waveguide.

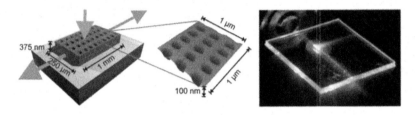

Figure 3.11. *Diagram of the 2D DFB laser [CHR 08]*

One-dimensional DFB lasers have been demonstrated with different types of organic materials [HEL 04, SPE 05, RAB 05]. The advantage of this structure is that we can easily change the resonance wavelength by changing the step of the lattice or the thickness of the material deposited (modification of the effective index).

The concept of the DBF laser can be extended to two-dimensional structures. These structures are made with 2D photonic crystals [CHR 08, DON 08, BAU 07]. It should be noted, though, that Christiansen *et al.* [CHR 08] obtained a lasing threshold of 470 nJ/mm^2 for emission at 595 nm.

The emission wavelength can be adjusted by modifying the step of the lattice. A post-fabrication adjustment of a few nanometers is possible by modifying the temperature of the component, notably by Peltier-effect devices. In 1971, Kögelnik and Shank were the first to design an organic DFB laser with rhodamine 6G [KÖG 71]. Other teams followed [MCG 98, RIE 00, KAR 07] – notably M.B. Christiansen *et al.* – who used a two-dimensional DFB cavity with organic materials and obtained a laser threshold of 47 µJ/cm² [CHR 08].

3.4.2.3. *DBR (distributed Bragg reflector) laser*

In DBR lasers, the active medium is placed between two Bragg lattices (see Figure 3.12). The DBR laser works on the principle of prohibited photon bands of the lattice surrounding the active medium. These cavities are similar to Fabry–Pérot cavities, but the advantage to them is that they show amplification only around the Bragg resonance. Thus, the narrow bandwidth of reflectivity prevents numerous longitudinal modes.

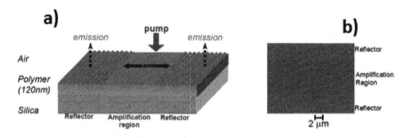

Figure 3.12. *Diagram of a DBR organic laser [VAS 06]*

The introduction of a cavity between two lattices creates a narrow transmission peak in the prohibited photonic band located in the fault [VAS 06]. The number of modes and their position can be controlled by changing the parameters of the lattice and the size of the cavity. The advantage of this type of laser is that the gain medium is located in an unstructured zone, which limits diffusion losses. Vasdekis *et al.* reported a threshold of 9.7 $\mu J/cm^2$ with a quality factor of 1000.

3.4.2.4. *2D photonic crystal laser*

2D photonic crystal lasers are two-dimensional periodic lattices with a high index contrast, with a cavity at its center, which is also known as a defect (Figure 3.13). The laser emission takes place perpendicularly to the plane. This type of laser facilitates good confinement of the photons, a low modal volume and the making of low-threshold lasers. At present, though they are perfectly functional, 2D photonic crystal lasers are still at the research stage, notably because of the high manufacture cost. Indeed, the techniques used, such as electron-beam lithography, make their manufacture a waste of time. In 2012, F. Gourdon *et al.* managed to make an optically pumped 2D photonic crystal laser, with a lasing threshold of 9.7 $\mu J/cm^2$ [GOU 12].

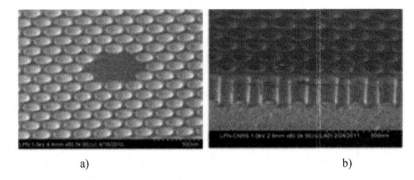

a)   b)

Figure 3.13. *2D photonic crystal laser (LPL/LPN):*
bird's eye view (a) and cross-section (b) [GOU 12]

3.4.2.5. *Fabry–Pérot (vertical) microcavities*

Fabry–Pérot microcavities are made of multilayer planar mirrors, between which is the gain medium. This type of laser is often less expensive, given the better-controlled, faster fabrication process. Among the various research projects conducted on such organic microcavities [MAS 00, JUN 01, MAR 05, SAK 08, RAB 09], M. Koschorreck *et al.* succeeded in obtaining a quality factor of 4500, with a threshold of 20 µJ/cm² with optical pumping [KOS 05].

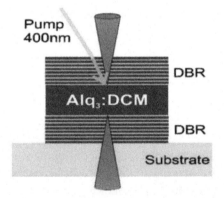

Figure 3.14. *Diagram of an optically pumped*
Fabry–Pérot cavity laser [KOS 05]

3.4.2.6. *Micropillar lasers*

Micropillar lasers emit from their surface, like Fabry–Pérot cavities. In addition to benefitting from excellent longitudinal confinement because of dielectric mirrors, these cavities are etched laterally to confine the field transversely. Aside from the technological difficulties which this adds, these structures should, in theory, facilitate a high quality factor and a low threshold, due to a greatly reduced modal volume. A.M. Adawi *et al.* managed to design a micropillar cavity containing organic material and exhibiting a quality factor of 680 [ADA 08].

Figure 3.15. *SEM photo of a micro-ring laser [ADA 08]*

3.5. A step toward the electrically pumped organic laser

3.5.1. *State of the art*

Of the various groups working on organic diode lasers, some have studied and created different approaches combining electrodes for electrical excitation and a laser cavity. These include P. Andrew *et al.* and M. Reufer *et al.* [REU 04], who successfully integrated a metal electrode whilst preserving the optically pumped laser effect. P. Andrew and his team used a DFB cavity and compared the threshold as a function of the presence or absence or a silver cathode [AND 02] (Figure 3.15). They obtained a threshold of 16 µJ/cm² without the metal present, whilst the addition of the layer of silver increases the threshold to 244 mJ/cm², corresponding to pumping equivalent to 2.4×10^7 A/cm² with electric pumping. This high threshold offers a good illustration of the very great absorption caused by metal electrodes.

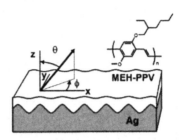

Figure 3.16. *Diagrammatic representation of the silver-covered DFB structure [AND 02]*

Using a similar approach, M. Reufer and his team [REU 04] attempt to reduce the absorption of the cathode by evaporating a thinner layer of silver than in P. Andrew's study – only 150 nm thick. He also added a 20 nm layer of ITO on the side of the anode, as shown by Figure 3.16.

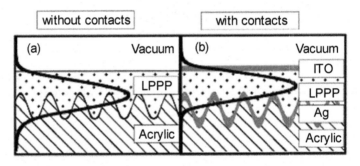

Figure 3.17. *DFB structure with a fine layer of silver [REU 04]*

According to the results of this study, they obtain a laser threshold at 6 μJ/cm^2 without ITO and without silver, but adding these two elements increases the threshold to only 7.6 μJ/cm^2, which is 758 A/cm^2 in equivalent electrical pumping. These results are interesting, because they suggest that the avenue of thin-layer metal electrodes can represent a favorable compromise from the electrical and optical point of view. One of the points of that compromise pertains to the thickness of ITO. Given the planar configuration of DFB structures, the resonant wave is liable to be guided in ITO if the latter is thick enough, which can increase the lasing threshold. For this reason, they reduce the thickness of ITO to only 20 nm, which constitutes a critical parameter limiting the injection of high current density.

Using another approach, M.C. Gwinner *et al.* [GWI 09] managed to couple a DFB cavity containing organic materials, with an organic field-effect transistor (OFET – see Figure 3.17).

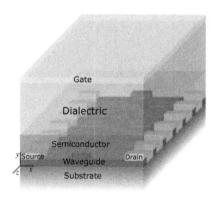

Figure 3.18. *OFET ("Organic Field-Effect Transistor") in a DFB cavity [GWI 09]*

In the absence of an electrode, they observe a lasing threshold at 4.1 µJ/cm² with optical pumping, whilst the complete configuration shown in Figure 3.17 enabled them to obtain a threshold at 4.6 µJ/cm², equivalent to a threshold of 460 A/cm² in electrical pumping. They also managed to make that structure work with electrical pumping – up to 100 V in the pulsed regime. However, no quality factor is mentioned in that article.

Now let us consider OLEDs in which steady-state wave effects and resonance between the electrodes have been studied.

A vertical microcavity OLED strategy was employed by F. Jean *et al.* [JEA 02]. For this purpose, they used an OLED whose cavity is formed of a layer of 100 nm of aluminum on the one hand, and 30 nm of aluminum on the other, forming a semi-transparent cathode. In order to control the emission wavelength of that microcavity, they varied the thickness of ITO from 200 nm to 370 nm, thereby modifying the size of the resonant cavity. With an optical excitation at 411 nm, they obtained significant spectral narrowing of the OLED, moving from a spectrum with an FWHM = 90 nm outside the cavity to an FWHM equal to 9 and 13 nm in the cavity, for a resonance wavelength equal to 555 nm and 530 nm, respectively (Figure 3.18).

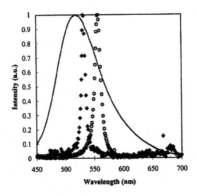

Figure 3.19. *Emission spectrum of the microcavity OLED [JEA 02]*

The work of S. Tokito *et al.* also pertains to cavity OLEDs, but for applications of spectral purity [TOK 99]. This is an OLED structure deposited on a multilayer dielectric mirror SiO_2/TiO_2. A steady-state wave is thus established between the dielectric mirror and the 180 nm aluminum cathode, and provokes resonance within the organic material, which spectrally refines its emission and therefore its spectral purity. Hence, it should be noted that a microcavity is made simply of a mirror and a metal electrode, even though the quality factor remains limited. In these publications, the variation of the thickness of the OLED between the two mirrors enables the authors to tune the resonance wavelength of the transmission peaks, with FWHM at 18 nm at $\lambda = 618$ nm, and up to FWHM = 10 nm at $\lambda = 462$ nm (Figure 3.19).

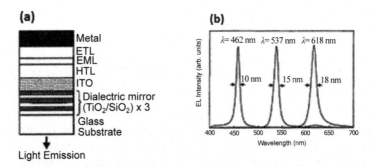

Figure 3.20. *a) Diagram of the microcavity OLED.*
b) Emission spectrum of the cavity [TOK 99]

These studies demonstrate the effects of interference in the organic layers. Thus, they highlight the importance of carefully designing the dimensions of the cavity.

A similar study conducted by X.J. Qiu *et al.* [QIU 06], devoted to improving the purity of the colors, demonstrates this type of optical resonance, with the one difference being that it uses a semi-transparent silver electrode 20 nm thick (Figure 3.20).

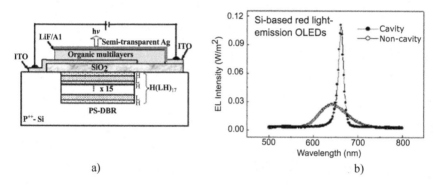

a)   b)

Figure 3.21. *a) Diagram of the microcavity OLED. b) Emission spectrum [QIU 06]*

These works show us a glimpse of an interesting prospect: indeed, it is possible to supplement the optical confinement offered by these cavities, with an upper dielectric mirror, superposed on the semi-transparent metal cathode. While such semi-transparent cathodes offer the advantage of reduced absorption, a question does remain: it pertains to the current density that these structures, with semi-transparent electrodes, can withstand – a few tens of nanometers.

By applying this strategy, X. Liu and team, in 2009, claimed the first demonstration of a laser effect by placing an OLED in a vertical microcavity [LIU 09]. In this study, they used an OLED comprising a 20 nm aluminum cathode, and a 71 nm anode of IVO ("Indium Vanadium Oxide"), all between two dielectric mirrors, composed of alternating layers of SiO_2/TiO_2 and ZnS/MgF_2 for the low and upper mirrors, respectively (Figure 3.21).

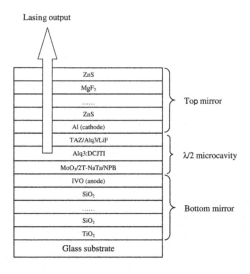

Figure 3.22. *Structure of the OLED in a vertical microcavity [LIU 09]*

This claim about the laser effect was roundly contested by the scientific community [SAM 07], in view of the weakness of the evidence offered. Indeed, they claim electrically pumped laser emission with a surprisingly low threshold of only 860 mA/cm², with a quality factor of $Q = 167$. The article also suffers a lack of precision as regards the type of electrical excitation. Nevertheless, the proposed structure is undeniably interesting, because it is able to combine the electrical excitation and the cavity, and paves the way for cavities with a high quality factor when the structure can be optimized.

Another argument in favor of this type of Fabry–Pérot cavity relates to optical confinement, which is perpendicular to the plane of the layers. Indeed, as we saw earlier, in the planar configurations put forward by P. Andrew and M. Reufer, the resonant wave is particularly exposed to cathodes along the reference axis, which leads to non-negligible cumulative lineic absorption. Because of this vertical disposition of the cavity, the structure envisaged by X. Liu is able to eliminate the guided waves in the electrodes, and thus decrease absorption.

3.5.2. *A step toward the organic laser diode (electrically pumped)*

Organic materials have a high photoluminescence (PL) quantum yield, a large effective cross-section of gain, and broad fluorescence which covers almost the whole of the visible spectrum. They can be easily used and offer low-cost manufacturing techniques. In spite of the fact that the first solid-state optically pumped organic laser was demonstrated in 1996 [TES 96], there has not yet been any demonstration of a working organic diode laser. This necessitates the design of a laser structure with a current density at the threshold laser at a level compatible with the current density of an organic light-emitting diode (OLED) – 100 A/cm^2 in alternating current (AC) and 0.1 A/cm^2 in direct current (DC) [FRO 97].

By analyzing multiple results published by numerous authors on optically pumped organic lasers, we propose an empirical approach which provides an estimation of the minimum quality factor necessary for the realization of an organic laser.

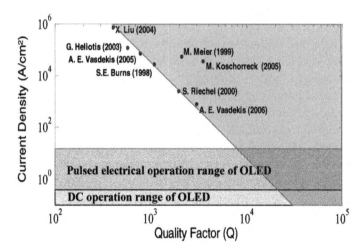

Figure 3.23. *Graphic representation of the different laser experiments with organic materials reported in the existing body of literature. (Current density equivalent to the excitation density obtained by optical pumping as a function of the quality factor)*

For this purpose, Figure 3.22 presents the density of excitation at the lasing threshold as a function of the quality factor for several experiments reported in the literature. The density of excitation converted into equivalent current density J_{TH} calculated using the following expression [CHA 11]:

$$J_{th} = \frac{q}{\chi h \upsilon \tau} I_{th}$$

[3.39]

where q is the charge of an electron, hv is the energy of the photons used in the experiments, τ is the lifetime of the excitons and I_{TH} is the lasing threshold when subjected to optical excitation. To take account of the radiative states, representing 25% of excited states, we introduce a factor $\chi = 0.25$. This calculation is only intended to provide an estimate.

In addition, this study facilitates good estimation of the pump energy necessary to obtain a laser effect by electrical pumping. Following on with the trend, to develop an electrically pumped organic laser, with a reasonable current density, we need to achieve a quality factor of $Q \sim 3 \times 10^4$ in DC and $Q \sim 10^4$ in AC. From this trend, we can deduce a strategy whereby structures are optimized in order to achieve this high quality factor.

Direct electrical pumping of a diode laser with an organic semiconductor continues to be a major problem. This is a very attractive objective which would facilitate the realization of simple, low-cost lasers emitting in the whole of the visible spectrum. New developments in the field of organic lasers have come to light to achieve this result. These developments include indirect electrical pumping, where a light source under electrical excitation is used to optically pump an organic semiconductor. In this approach, the charge carriers are not directly injected into the active medium, but into a diode laser or an LED (based on an inorganic semiconductor) which, in turn, optically pumps the material. In addition, the compromises listed above for electrical pumping of the organic material are eliminated, because it is pumped optically. Riedl demonstrated a DFB organic laser pumped by an inorganic diode laser emitting in the violet [RIE 06]. Multiple demonstrations of organic lasers, based on conjugate polymers, pumped by a diode, have recently been published [SAK 08, VAS 06]. It is possible to go further in the way of simplification and miniaturization. For example, Yang made an organic semiconductor laser pumped by an inorganic LED [YAN 08].

Since 2000, numerous works have been published, demonstrating and studying the obstacles to the realization of an organic diode laser pumped electrically [BAL 02, LIS 01]. Generally speaking, these problems can be classified into two categories:

1) Optical problems:

Optical problems are generally linked to:

– low refractive indices of organic materials (n = 1.7). The confinement of the light inside the OLED is limited;

– absorption of the photons by the metal electrodes [TES 99]. This is particularly important in organic semiconductor lasers where the resonator is placed in the plane of the emissive layer, thus leading to guided modes which extend to the electrodes;

– other phenomena of non-radiative losses. In this case, we are dealing with exciton–exciton annihilations: two excited states lead to a single excited state and a ground state (S_0), without emission. Beyond a certain threshold of concentration ($>10^{17}$ cm^{-3}), the density of excited states reaches saturation. Thus, at least half of the excitons are lost. The quenching of the singlet excitons by the triplet excitons is also a major source of losses standing in the way of the successful design of the organic diode laser.

2) Electrical problems:

The main electrical problems encountered are, notably, due to the presence of polarons: electrical excitation involves the injection of charges, which creates polarons. These are sources of additional absorption losses and can reduce the number of radiative excitons by phenomena of quenching of the excitons. Generally, the polaron absorption spectrum is very broad and covers a wide part of the visible spectrum.

However, the major problem with an organic laser remains the low current density achieved by electrically pumped OLEDs. The lowest threshold current densities were obtained by Kozlov, near to 100 A.cm^{-2} in a DFB laser structure with an organic layer doped with DCM:Alq3 [KOZ 00]. Current densities verging on 1 kA.cm^{-2} were achieved in a pulsed OLED regime [TES 98] and in a field-effect organic diode [TAK 08].

A possible source of non-radiative losses is created by the metal electrodes. In the configuration of the organic diode, the resonator is in the plan with the electrodes, which absorb light. We can suppose that the energy density at the threshold is greatly increased when a metal electrode is present; this observation was demonstrated with optical pumping [AND 02].

There is a certain compromise to be found between the optical and electrical properties. Indeed, it is possible to improve the design to reduce these losses, using two transparent electrodes [LAT 06] and/or optimizing the structure [CHA 11].

Therefore, it is necessary to develop structures that are capable of mitigating these negative effects. The organic semiconductor must have high mobility of the charge carriers, a high optical gain and be incorporated into an optical resonator with very low losses. This approach opens the way to new research on organic–inorganic photonic components such as the use of photonic crystal microcavities. In this approach, the use of the prohibited photonic bands in a 2D photonic crystal helps to improve the coupling of the light with the resonant mode. Another approach is to improve the optical and electrical performances of OLEDs using the plasmonic effect. These properties are discussed in the next chapter.

3.6. Conclusion

In this chapter, we have recapped the principles of the laser effect, which involves the pumping of a gain medium and the use of a resonant cavity, which acts as a filter and an amplifier of the radiation obtained. Then, we looked at the peculiarities inherent to organic materials. Because the electrically pumped organic laser diode has yet to be demonstrated, we presented the current state of the art on optically pumped organic lasers. To conclude the chapter, we examined the problems posed by the quest for the electrically pumped organic laser and the possible ways to achieve this objective.

4

Organic Plasmonics:
Toward Organic Nanolasers

The LSP (Localized Surface Plasmon) is one of the more interesting optical properties of metal nanostructures. It results from the collective oscillation of the electron cloud at the surface of a metal nanoparticle (NP). Thus, the electromagnetic (EM) field produced in the immediate vicinity of the NP may surpass the excitation field by several orders of magnitude. This property is used for various applications – in particular, for Metal-Enhanced Fluorescence (MEF).

In this chapter, we shall introduce the fundamental principles of the surface plasmon. We describe the optical response of metals, presenting the different models which describe their dielectric constants. Then, we examine the effects of the plasmon on the properties of an emitter placed in proximity to a metal NP. Finally, there is a presentation of the effects of LSP on the optical and electrical performances of organic emitters – OLEDs in particular – with a view to using this effect to encourage laser emission.

It should be noted that this chapter is essentially based on Samira Khadir's doctoral thesis, defended in 2016 and supervised by part of the team of authors [KHA 16].

4.1. Optical properties of metals

Generally, the propagation of an EM wave in a given medium is described by the Maxwell equations:

$$div \, \vec{D} = \rho_{free} \hspace{4cm} [4.1]$$

$$div \, \vec{B} = 0 \hspace{4cm} [4.2]$$

$$\overrightarrow{rot} \, \vec{E} = -\frac{\partial \vec{B}}{\partial t} \hspace{4cm} [4.3]$$

$$\overrightarrow{rot} \, \vec{H} = \vec{J}_{free} + \frac{\partial \vec{D}}{\partial t} \hspace{3cm} [4.4]$$

These equations link the electric field $\vec{E}$, magnetic field $\vec{H}$ and the magnetic induction vector $\vec{B}$ and electric induction vector $\vec{D}$. They also involve the free current density $\vec{J}_{free}$ and the free charge density ρ_{free}.

The electrical and magnetic field vectors are linked to the electrical and magnetic induction vectors by the following equations:

$$\vec{D} = \varepsilon \, \vec{E} \hspace{4cm} [4.5]$$

$$\vec{B} = \mu \, \vec{H} \hspace{4cm} [4.6]$$

ε and μ are, respectively, the electric permittivity and magnetic permeability.

They represent the polarization response of the medium subjected to the electrical field $\vec{E}$ and the magnetization response of that same medium subjected to the magnetic field $\vec{H}$.

The optical response of a metal is described by the interaction between the solid, represented by its dielectric function $\varepsilon(\omega)$, and light, represented by an EM wave. Various models have been put forward to represent the dielectric function of metals. The model for each case is chosen on the basis of the nature and properties of the metal under study.

4.1.1. *Drude's model*

The first model used to study the behavior of the conduction electrons in a metal was that developed by Drude [DRU 00]. In this ideal model, the metal is considered a gas of free electrons (conduction electrons) initially ignoring the lattice (concept of plasma). The dielectric function in this case is given by:

$$\varepsilon_D = \varepsilon_\infty - \frac{\omega_p^2}{\omega^2 + i\omega\gamma}$$
[4.7]

Its complex form can be written as follows:

$$\varepsilon_{metal} = \varepsilon_{real} + i\varepsilon_{imaginary} = \varepsilon_\infty - \frac{\omega_p^2}{\omega^2 + \gamma^2} + i\frac{\omega_p^2 \gamma}{\omega(\omega^2 - \gamma^2)}$$
[4.8]

where γ represents a damping coefficient due to the collisions between the electrons, and between electrons and phonons and other impurities and faults.

ω_p is the plasma frequency of the metal, given by the following relation:

$$\omega_p = \sqrt{\frac{n\,e^2}{\varepsilon_0 m}}$$
[4.9]

Drude's model only takes account of intra-band electron transitions, which is a poor approximation in the case of noble metals at visible–near UV frequencies, as illustrated in Figure 4.1, which shows a comparison of the real and imaginary parts of the dielectric function, calculated using Drude's model, and those measured experimentally. We can see a huge difference, particularly in terms of the imaginary part, and especially at low wavelengths.

4.1.2. Drude–Lorentz model

Transitions of the electrons from the filled valence band to the conduction band can considerably alter the optical properties of metals. In the case of alkaline metals, these transitions only occur at high frequencies and have very little influence on the dielectric function in the optical domain. Hence, such metals are accurately represented by Drude's model. On the other hand, in the case of noble metals, a correction must be made in order to take account of the inter-band transitions which take place between the valence band d and the conduction band s-p.

This correction is introduced in the context of the so-called Drude–Lorentz model [JOH 72]. The dielectric function is therefore written in the following form:

$$\varepsilon_{DL}(\omega) = \varepsilon_D(\omega) + \varepsilon_L(\omega) \qquad [4.10]$$

The estimation of $\varepsilon_{DL}(\omega)$ deems that the valence electrons are damped, forced harmonic oscillators. In the case of a single oscillator, $\varepsilon_{DL}(\omega)$ is given by the equation:

$$\varepsilon_{DL}(\omega) = \varepsilon_\infty - \frac{\omega_p^2}{\omega^2 + i\omega\gamma} - \frac{\Delta\varepsilon.\omega_L^2}{(\omega^2 - \Omega_L^2) + i\Gamma_L\omega} \qquad [4.11]$$

Ω_L and Γ_L, respectively, represent the strength of the oscillator and the spectral breadth of the Lorentz oscillators. $\Delta\varepsilon$ can be interpreted as a weighting factor ω_L.

The real and imaginary parts of that function are plotted in Figure 4.1. We can see better agreement with the experimental values obtained by Johnson and Christy [JOH 72] across the whole of the spectrum.

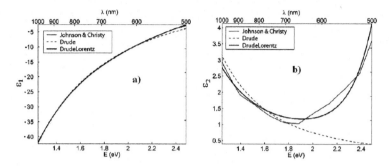

Figure 4.1. *Dielectric function of gold measured experimentally by Johnson and Christy [JOH 72] and adjusted by the Drude and Drude–Lorentz models. a) real part; b) imaginary part. From [VIA 05]*

4.1.3. *Drude's model with two critical points*

The Drude–Lorentz model offers a sufficiently accurate description of the optical properties of a number of noble metals such as silver. In the case of gold, because of its optical properties that are difficult to analytically represent in the visible spectral range (inter-band transitions are significant in that range), this model does not give good results, even when we include several Lorentzian terms [NOV 06]. We need to move onto a model where the contribution of these inter-band transitions are taken into account more completely, essentially in the optical range. Etchegoin *et al.* [ETC 06, ETC 06] proposed a model whose analytical formula is as follows:

$$\varepsilon_{\text{D2CP}}(\omega) = \varepsilon_{\infty} - \frac{\omega_p^2}{\omega^2 - i\omega\gamma} + \sum_{p=1}^{p=2} G_p(\omega) \qquad [4.12]$$

where:

$$G_p(\omega) = A_p\,\Omega_p \left(\frac{e^{i\varnothing_p}}{\Omega_p - \omega - i\Gamma_p} + \frac{e^{-i\varnothing_p}}{\Omega_p + \omega + i\Gamma_p} \right) \qquad [4.13]$$

The first two terms in equation [4.12] represent Drude's classic contribution. The sum represents the contribution of the inter-band transitions with the amplitude A_p, the gap energy Ω_p, the phase $\varnothing_p$ and the spreading Γ_p.

The comparison of the adjustment of that function against the four-Lorentzian (4L) model [FAO 07] is shown in Figure 4.2. It shows the precision of Drude's model with two critical points in the description of the dielectric functions of gold and silver, with fewer parameters to determine.

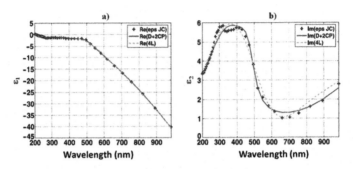

Figure 4.2. *Dielectric function measured experimentally by Johnson and Christy, compared to the 4L model and Drude's model with two critical points. a) Real part; b) imaginary part. From [VIA 08]*

4.2. What is a plasmon?

There are three types of plasmon: the volume plasmon, the delocalized surface plasmon and the localized surface plasmon.

4.2.1. *Volume plasmon*

A metal can be assimilated to a plasma in which the conduction electrons are moving freely. The electron density can fluctuate, and the oscillation energy of the electrons can be quantified. The energy quantum associated with these oscillations gives us the volume plasmon whose pulsation is given by:

$$\omega_p = \sqrt{\frac{n\,e^2}{\varepsilon_0 m}} \qquad [4.14]$$

The energy of these plasmons is of the order of 10–20 eV. Therefore, it is not possible to excite them with an optical excitation in the visible – hence the need to use electrons or X rays to excite them.

4.2.2. Delocalized surface plasmon

At the metal–dielectric interface, the Maxwell equations show that electromagnetic waves may propagate. Those waves are associated with oscillations of the plasma of free electrons at the surface of the metal (Figure 4.3(a)). They are known as "delocalized surface plasmons". The maximum intensity of the field associated with that wave occurs at the metal–dielectric interface. The intensity of the EM field decreases exponentially in both media in the directions perpendicular to the interface (Figure 4.3(b)).

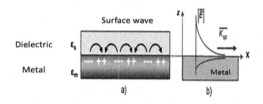

Figure 4.3. *a) Diagrammatic representation of the oscillation of charges at the metal–dielectric interface; b) Exponential decrease in the EM field on both sides of the interface*

Solving the Maxwell equations whilst applying the conditions of continuity at the metal–dielectric interface yields the following dispersion relation [RAE 86]:

$$k_x = \frac{\omega}{c}\left(\frac{\varepsilon_d \varepsilon_m}{\varepsilon_d + \varepsilon_m}\right)^{\frac{1}{2}} \qquad\qquad [4.15]$$

ε_m and ε_d are, respectively, the dielectric constants of the metal and the dielectric.

The dispersion curve for the surface plasmon is always situated below that of the light cone as indicated in Figure 4.4. This means that it is not possible to achieve coupling between the radiative light and the surface plasmon by directly lighting the metal with a light source. To make this coupling possible, there are a variety of techniques that can be used, such as: grating coupling and prism coupling [RAE 86].

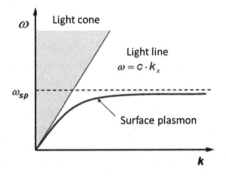

Figure 4.4. *Dispersion curve for a surface plasmon and light line in a vacuum*

4.2.3. *Localized surface plasmon*

Surface plasmons also exist for metallic nanoparticles. In this case, the effect of the size and composition of the nanoparticles and their environments have a marked influence on the position of resonance. These plasmons may be radiative and be coupled directly to the incident light. They exhibit significant confinement of the EM field in the vicinity of the metallic nanoparticles and they only propagate over nanometric distances. That is why they are called "localized surface plasmons" (LSPs).

The LSP resonance is the result of the interaction of light with a metallic nanoparticle. It can be qualitatively described by a simple approach: a metallic nanoparticle is made up of an ionic core and freely moving electrons. When that NP is illuminated by an EM field, an opposing field is created within it, causing an oscillation of the charges, whose intensity is maximal for the resonant frequency, which can be determined by measuring the quenching spectrum which shows the existence of the LSP.

For small metallic particles, unlike in the case of bulk metals, the external EM field can penetrate to the interior of the nanoparticle and cause the displacement of the conduction electrons in relation to the metal ions, inducing a local field opposite to the excitation field (see Figure 4.5). The coherent motion of the electrons and the induced field form an oscillator, whose behavior is defined by the mass and effective charge of the electrons, the electron density and the geometry of the particle. The resonances generated by this oscillator are known as LSPRs (Localized Surface Plasmon Resonances).

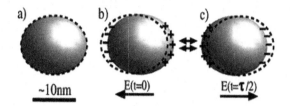

Figure 4.5. *Representative diagram of the oscillation of charges in a metallic nanoparticle in response to an external EM field*

The majority of physical effects associated with the localized surface plasmon can be qualitatively represented by the basic model of an oscillator. The LSP resonances are situated in the visible spectral domain, also extending into the near infrared, and depend heavily on the nature of the metal, the form, size and environment of the particles. When the NP is excited at the resonance wavelength, the amplitude of the EM field generated in the vicinity of the NP may be greater than the excitation field by several orders of magnitude (enhancement of the near-field optics). Similarly to a conventional oscillator, the damping coefficient of the electrons limits the maximum amplitude of the resonance and determines its spectral width.

In addition, metallic nanoparticles generally have more than one mode of oscillation. These modes have different charge distributions and field distributions. For the lowest mode, the distributions are dominated by a dipolar nature. The higher energy modes can be associated with multipolar higher-order charge distributions.

4.3. Theoretical approach to the localized surface plasmon (LSP)

There are multiple approaches to study the localized surface plasmon, theoretically and/or numerically. They depend primarily on the conditions linked to the solving of the Maxwell equations and, in a way, on the geometry used. In the following, we shall recap the basic principles of the most commonly used methods.

4.3.1. Mie's theory

To accurately calculate the LSP, we must solve the Maxwell equations for the NPs, using the appropriate boundary conditions. The analytical solution can only be obtained with certain geometrical forms. It is to Gustav Mie [MIE 08] that we owe the development of a theory which gives the exact solution for spherical NPs, providing they are sufficiently far apart (no interaction between them). In this case, the effective quenching cross-section is given by:

$$\sigma_{ext} = \sigma_{abs} + \sigma_{diff} = \frac{2\pi}{k^2} \sum_{L=1}^{\infty}(2L + 1) \, Re \, [a_L + b_L] \qquad [4.16]$$

where:

$$a_L(x) = \frac{m\psi_L(mx).\psi'_L(x) - \psi_L(x).\psi'_L(mx)}{m\psi_L(mx).\eta'_L(x) - \psi'_L(mx).\eta_L(x)} \qquad [4.17]$$

$$b_L(x) = \frac{\psi_L(mx).\psi'_L(x) - m\psi_L(x).\psi'_L(mx)}{\psi_L(mx).\eta'_L(x) - m\psi'_L(mx).\eta_L(x)} \qquad [4.18]$$

$x = k.R$, where k is the wave vector of light in the transparent host, and R is the parameter of the size of the nanoparticle. $m = \dfrac{n_m}{n}$, where n_m and n represent, respectively, the refractive indices of the metal and of the surrounding dielectric host matrix. ψ_L and η_L are the Bessel–Riccati functions. These equations are difficult to solve analytically, but can be computed numerically.

4.3.2. Dipolar model or quasi-static approximation

The quasi-static approximation is justified for spherical NPs whose diameters are much smaller than the resonance wavelength of the LSP ($2R < \lambda / 10$).

Indeed, the electrical field can be considered to be uniform within the NP, which can therefore be described by a dipole (Figure 4.6(a)).

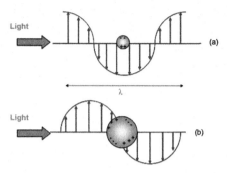

Figure 4.6. *Representation of the charge distribution in the spherical NP under the influence of an EM field. a) Very small NP in comparison to the LSPR wavelength; b) NP comparable to the LSPR wavelength*

In the context of this approximation, the effective cross-sections of quenching and diffusion are given, respectively, by [BOH 83]:

$$\sigma_{diff} = \frac{144V^2\pi^4\varepsilon_D^2}{\lambda^4}\frac{(\varepsilon_1-\varepsilon_D)^2+\varepsilon_2^2}{(\varepsilon_1+2\varepsilon_D)^2+\varepsilon_2^2} \qquad [4.19]$$

$$\sigma_{ext} = \frac{18V\pi\varepsilon_D^{3/2}}{\lambda}\frac{\varepsilon_2}{(\varepsilon_1+2\varepsilon_D)^2+\varepsilon_2^2} \qquad [4.20]$$

ε_D is the dielectric constant of the surrounding host material. ε_1 and ε_2 are, respectively, the real and imaginary parts of the metal's dielectric function. For a given environment, the ratio between the two effective cross-sections is proportional to the volume of the nanoparticles:

$$\frac{\sigma_{diff}}{\sigma_{ext}} \propto \frac{V}{\lambda^3} \qquad [4.21]$$

For nanoparticles of small dimensions, diffusion is therefore negligible and quenching is largely dominated by absorption. When the size of the NPs increases, this approximation reaches its limits, because the field inside the NP can no longer be considered uniform (Figure 4.6(b)). In this case, multipolar modes emerge, and their contributions are taken into account by Mie's theory.

4.3.3. *Theories on effective dielectric functions*

Another approach for determining the optical properties of metallic NPs is to replace the heterogeneous system, formed by the NPs and the surrounding dielectric medium, with a homogeneous medium with an effective dielectric function ($\varepsilon_{eff1} + i\varepsilon_{eff2}$). Multiple theories based on this approach have been developed. The first one, put forward by Newton [KRE 95], defines the effective dielectric function as the mean of the dielectric functions of the NPs and the dielectric medium, weighted by their volumetric fraction of inclusion f:

$$\varepsilon_{eff} = f\varepsilon + (1 - f)\varepsilon_D \qquad [4.22]$$

However, the best approach in terms of these theories is the Maxwell–Garnett approach, whereby the heterogeneous system should be replaced with a homogeneous material exhibiting the same dielectric polarization when it is subjected to excitation by light [LIC 26, MAX 04]. In this case, the effective dielectric function is given by:

$$\frac{\varepsilon_{eff} - \varepsilon_D}{\varepsilon_{eff} + 2\varepsilon_D} \qquad [4.23]$$

The result yielded by this theory for NPs of extremely small dimensions accords very closely to that found by the dipolar approximation. The Maxwell–Garnett model exhibits an interesting advantage, as it can be modified to take account of the interactions between NPs and of their non-spherical form.

4.3.4. *Numerical study by FDTD (Finite-Difference Time-Domain)*

In order to get around the limitations of theoretical models, it is often necessary to conduct a numerical study using the Finite-Difference Time-Domain (FDTD) method, on a plasmonic device made of metallic nanoparticles deposited on a dielectric substrate. There are a host of readily available references on the principles of this method.

Remember that the FDTD is a method for solving partial differential equations. It was put forward by Yee in 1966 [YEE 66]. Thanks to its advantages and the high-performing software, this method is becoming increasingly widely used for applications in various domains. The FDTD method is capable of solving the Maxwell equations describing situations where the variation, over time, of the electrical field $\vec{E}$ (or of the magnetic field $\vec{H}$) is dependent on the spatial variation of the magnetic field $\vec{H}$ (or electrical field $\vec{E}$). It consists of solving this system of partial differential equations by discretization with respect to both time and space. Indeed, the differential of a function f is calculated as follows:

$$f'(x) = \lim_{\Delta \to o} \frac{f(x+\Delta) - f(x-\Delta)}{2\Delta} \qquad [4.24]$$

$$f'(x) \approx \frac{f(x+\Delta) - f(x-\Delta)}{2\Delta} \qquad [4.25]$$

The FDTD method is based on the principle of centered-difference discretization of the partial differentials appearing in the Maxwell equations. In space and in one of the primary directions (x, y or z), each calculation point is separated from its neighbor by a distance Δ, known as the step of spatial discretization. The computation volume, therefore, will be a rectangular parallelepiped, divided into ($N_x \times N_y \times N_z$) elementary cells, each of the volume ($\Delta x \times \Delta y \times \Delta z$). Δx, Δy and Δz respectively denote the spatial discretization steps in the directions Ox, Oy and Oz .

Finally, to ensure the stability of the FDTD method, the spatiotemporal discretization steps must satisfy the following criterion [GÉR 98]:

$$\Delta t \leq \left[c. \sqrt{\frac{1}{\Delta x^2} + \frac{1}{\Delta y^2} + \frac{1}{\Delta z^2}} \right]^{-1} \qquad [4.26]$$

4.4. Parameters influencing the localized surface plasmon

One of the most attractive aspects of the localized surface plasmon in the various applications is the sensitivity of its resonance to numerous parameters. The possibility of controlling these factors gives us the ability to manipulate light on a nanometric scale, thus opening the way for new applications.

4.4.1. *Effect of size*

The size of the metallic nanoparticles has a considerable effect on the peak of surface plasmon resonance. In this context, we can distinguish two regimes. First, we have the case of small nanoparticles ($R \sim 50\ nm$), where the size mainly affects the intensity and the width of the resonance peak, whereas its effect on the position of the peak is reduced. In the case of large nanoparticles ($R > 50\ nm$), the effect of size is apparent in terms of the position of resonances [GAR 11].

4.4.2. *Effect of form*

Another parameter influencing the efficiency of quenching of the nanoparticle is its geometric form. Even a slight morphological deviation from a spherical form has a considerable impact on the LSPR. This is illustrated in Figure 4.7, showing the quenching spectra measured in the case of an NP of Ag having three different forms (spherical, pentagonal and triangular) [MOC 02].

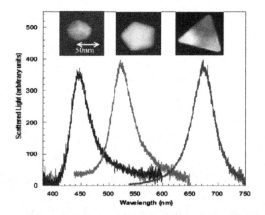

Figure 4.7. *Quenching spectra measured for silver NPs having different forms. From [MOC 02]*

4.4.3. *Effect of composition*

Figure 4.8 represents the effective quenching cross-section of spherical particles of copper, gold and silver of the same volume, calculated using the quasi-static approximation. From this figure, we can deduce that the nature of the metal of which the nanoparticle is made affects the position of the resonance peak, its amplitude (the LSPR is more marked in the case of silver than those of gold and copper) and its width (the resonance peak is narrower with silver than it is with gold and copper). Hence, an effect is produced on the lifetime of the surface plasmon τ_{pl} (which is inversely proportional to the full width at half maximum (FWHM) of the peak $\Delta\omega\,\tau_{pl} = 1$). The effect observed is greater in the case of silver.

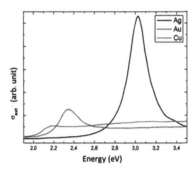

Figure 4.8. *Effective quenching cross-section of spherical nanoparticles of noble metals of the same volume in a silica host material, calculated using the quasi-static approximation. From [HAM 12]. For a color version of this figure, see www.iste.co.uk/boudrioua/lasers.zip*

4.4.4. *Effect of environment*

The immediate environment of the metallic nanoparticle has a significant influence on the LSPR. Figure 4.9 represents the effective quenching cross-section of a spherical nanoparticle of gold in different dielectric host matrices. We can see a red-shift of the LSPR, accompanied by an increase in its amplitude and the narrowing of its spectral width (and hence an increased lifetime of the surface plasmon) when the index of the dielectric medium increases.

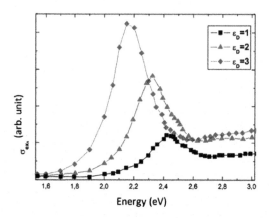

Figure 4.9. *Effective quenching cross-sections for nanoparticles of gold, 10nm in diameter for different host materials, calculated with the quasi-static approximation. From [GAR 11]. For a color version of this figure, see www.iste.co.uk/boudrioua/lasers.zip*

4.5. Plasmonic materials and their properties

The necessary condition to produce the plasmonic effect is the presence of a high density of free charges, resulting in a negative dielectric constant. However, the oscillation of these free charges also leads to energy dissipation by absorption, which can affect the overall energy balance. Consequently, a compromise must be found between these two effects, essentially by choosing the right material to use.

Owing to their high conductivity, metals are prime candidates for applications in plasmonics. Silver and gold are the most commonly used, thanks to their relatively low absorption in the visible and near-IR range (see Figure 4.10). Silver is the best conductor, exhibiting the lowest degree of absorption. However, it is chemically unstable in comparison to gold. The latter is the best material, after silver, in terms of losses, but is, of course, more expensive. Copper, which is much more cheaply available, exhibits properties similar to those of gold in the visible range, making it a promising candidate to replace gold and silver. However, because of its chemical instability, it is very difficult to manufacture copper-based devices. Aluminum, too, in spite of its high absorption in the near-IR domain, has recently given rise to numerous studies for applications in plasmonics. The real part of its dielectric function is negative in UV, up to wavelengths below

200 nm, whilst its imaginary part remains smaller than those of gold and silver, as indicated in Figure 4.10. Though other metals have been studied, their use is limited by their rather high losses in the visible range. For example, platinum and palladium have been used as plasmonic materials in systems where the material's catalytic activity is important for the operation of the complete device [TOB 01].

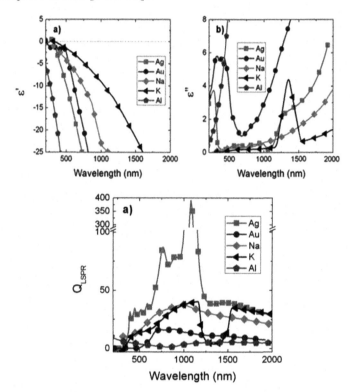

Figure 4.10. *Quality factor of the localized surface plasmon of a number of metals. From [WES 10]. For a color version of this figure, see www.iste.co.uk/boudrioua/lasers.zip*

Figure 4.10(b) shows the localized surface plasmon quality factor calculated in the case of a spheroidal NP for various metals [WES 10]. This factor is proportional to the ratio between the real and imaginary parts of the dielectric function (Q $\propto$ ε'/ε). This figure clearly shows why silver is the chosen plasmonic material for many applications.

Note that by using alloys – for instance, alloys of noble metals and transition metals – it is possible to control the band structure by adjusting the proportion of each metal in the alloy. In an experiment reported in [BOB 09], it was shown that the gold–cadmium alloy forms a unique band structure, and shifts the absorption peak into a clearly defined spectral range. Thus, the losses are reduced across the rest of the spectrum. This is due to the modifications of the Fermi level induced by the doping of the gold with a very specific percentage (< 10%) of cadmium.

We can also point to a number of studies which look at the use of semiconductors. Given how easily they can be fabricated, and how readily we can control their properties such as the concentration of charge carriers, semiconductors are considered potential candidates for plasmonics. In order for them to be characterized as low-loss materials, the stop-band energy and the plasma frequency (charge density) of the semiconductor must be larger than the range of frequencies of interest. A high plasma frequency makes for negative real permittivity, whilst a wide stop-band minimizes losses due to inter-band transitions. Semiconductors need to be heavily doped with guest materials to give them cutoff frequencies near to the visible range. Doped indium-tin oxide (ITO) and zinc oxide (ZnO) exhibit advantageous plasmonic properties in the near-IR range. However, the use of semiconductors in the visible range is a challenge which has yet to be overcome.

Finally, in recent years, graphene has attracted a great deal of attention thanks to its unique band structure and its high charge mobility. It is a two-dimensional material whose charge density is of the order of 10^{11}–10^{13} cm^{-2}, which can be controlled electrically by application of a potential difference [LI 08]. Graphene can give rise to a plasmonic effect similar to the surface plasmon engendered over a metal–dielectric interface with a different dispersion relation [STE 67, HWA 07]. Its plasmonic performances in the THz range have already been reported in the existing body of literature [RYZ 07, DUB 09]. However, at NIR frequencies, the losses in graphene may be comparable to those of noble metals. This makes it less attractive as a plasmonic material for use at telecoms frequencies and visible frequencies.

4.6. Optical properties of an emitter in the vicinity of a metallic NP

The presence of metallic nanostructures in the vicinity of active nano-objects such as atoms, molecules and quantum dots leads to modifications of their optical properties. This is due, as previously highlighted, to the fact that these NPs support plasmon modes engendering an intense EM field, in their vicinity, which may be stronger than the excitation field. This phenomenon has been widely used for different applications such as SERS (Surface Enhanced Raman Spectroscopy), biomolecule detection, fluorescence enhancement and improvement of the performances of optoelectronic devices such as solar cells and LEDs.

In this section, we shall focus on understanding the process of interaction between an NP and an active object, by means of the theoretical model developed by Khurgin *et al.* [KHU 09]. In particular, we shall discuss the processes responsible for modifications of the absorption, photoluminescence and electroluminescence.

This model is based on the quasi-static approximation presented above, and the effective volume theory [MAI 06]. We consider the case of a spherical metallic NP with radius a placed in a dielectric medium having a constant ε_D, as described previously. This NP can support multipolar resonance modes where only the dipolar (order 1) mode is radiative, and can be outcoupled with the external field. On the other hand, the higher-order modes are non-radiative oscillations considered to be a source of absorption losses. Using the expression of the dipolar moment obtained in the previous section (written here as p_1), in the context of the quasi-static approximation, the radiative power of that dipole is written:

$$P_{rad} = \frac{n^3 \omega_1^4}{12\pi\varepsilon_0\varepsilon_D c^3} p_1^2$$

[4.27]

On the basis of this expression, we can deduce the radiative decay rate of the dipolar mode [KHU 09]:

$$\gamma_{rad} = \frac{2\omega_1}{3\varepsilon_D}\left(\frac{2\pi a}{\lambda_1}\right)^3 = \frac{2\omega_1}{3\varepsilon_D}\chi^3 \qquad [4.28]$$

where ω_1 represents the resonance frequency of the dipolar mode, λ_1 is the corresponding wavelength and $\chi = 2\pi a/\lambda_1$.

This mode also presents a nonradiative decay rate γ_{nrad} due to the imaginary part of the metal's dielectric function. This rate is comparable to the damping coefficient of the electrons in Drude's model γ. The total decay rate is therefore given by:

$$\gamma_l = \begin{cases} \gamma_{rad} + \gamma & \text{if } l = 1 \\ \gamma & \text{if } l \geq 2 \end{cases} \qquad [4.29]$$

Thus, only the dipolar mode gives rise to the radiative coupling efficiency, given by:

$$\eta_{dp} = \frac{\gamma_{rad}}{\gamma_{rad} + \gamma} = \frac{2Q\chi^3}{3\varepsilon_D + 2Q\chi^3} \qquad [4.30]$$

where the factor $Q = \dfrac{\omega_1}{\gamma}$.

This model will be applied to study the various optical properties of an active object placed in proximity to the NP, as we shall discuss below.

4.6.1. *Modification of absorption*

Absorption enhancement by plasmonic effect of a metallic NP is an important phenomenon for two reasons: the first pertains to its use to improve the absorption efficiency in certain applications such as solar cells [WES 00, RAN 04] and photodetectors [TAN 08, SCH 05]; the second is the fact that this phenomenon is part of the process of enhancement of photoluminescent emitters. An enhancement of the emitter's absorption leads to an improvement of its emission. To study this phenomenon, we consider a molecule in a dielectric medium in proximity to a metallic NP supporting plasmon modes, as explained above. When that system is excited by an EM wave in the presence of the NP, only the dipolar mode can be coupled to the incident wave, and the phenomenon of absorption in this case can be described in two stages: the first is a coupling of the incident light with the dipolar mode of the surface plasmon, and the second is that the energy from the dipolar mode is absorbed, with an effective absorption cross-section σ_a, by the molecule situated at a distance d from the metal surface. The decay rate of the dipolar mode due to absorption by the molecule is given by [KHU 09]:

$$\gamma_{abs} = \frac{c}{n} \frac{N_a \sigma_a}{V_{eff}} \left(\frac{a}{a+d} \right)^6 \qquad [4.31]$$

where n represents the refractive index of the medium, N_a is the number of molecules surrounding an NP and V_{eff} represents the effective volume of the plasmonic mode.

The total enhancement factor F_A of the molecule's absorption by the presence of the NP is given by the ratio between the electrical fields in the presence and in the absence of the NP, defined as follows:

$$F_A = \left| \frac{E_{max}}{E_0} \right|^2 \left(\frac{a}{a+d} \right)^6 \qquad [4.32]$$

$$F_A = \left(\frac{\omega_1}{\omega_{ex}} \right)^2 \frac{2}{\left[Q^{-1} + Q_a^{-1} + 2\chi^3 / 3\varepsilon_D \right]^2 + \delta_{ex}^2} \left(\frac{\chi}{\chi + \chi_D} \right)^6 \qquad [4.33]$$

where Q_a is the absorption factor given by:

$$Q_a = \frac{\omega_1}{\gamma_{abs}} \text{ and } \delta_{ex} = 2 \left(1 - \frac{\omega_{ex}}{\omega_1} \right) \qquad [4.34]$$

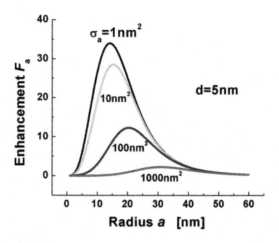

Figure 4.11. *Absorption enhancement factor as a function of the dimensions of the NP, and of the effective absorption cross-section. From [KHU 09]*

The absorption enhancement factor has been calculated for the case of a nanosphere of gold placed at a distance $d = 5$ nm from an absorbent molecule having different effective absorption cross-sections $N_a\sigma_a$. The result, as a function of the radius of the sphere, is shown in Figure 4.11. The enhancement is maximal at low effective absorption cross-sections; in other words, materials which are good absorbers initially cannot be enhanced by the presence of the NP.

4.6.2. *Modification of electroluminescence*

The phenomenon of electroluminescence enhancement in the presence of metallic NPs has been the subject of numerous experimental studies, with the main aim of demonstrating improvements in the performances of LEDs [YAN 09, YAN 09, FUJ 10]. The objective in this section is for readers to understand the phenomenon of enhancement of the electroluminescence of a molecule in the vicinity of a metallic NP from the theoretical standpoint.

In the absence of the metal, the molecule has a radiative decay rate $1/\tau_{rad}$ and a nonradiative decay rate $1/\tau_{nrad}$, and its radiative efficiency (or quantum yield) is given by:

$$\eta_{rad} = \frac{\tau_{rad}^{-1}}{\tau_{rad}^{-1} + \tau_{nrad}^{-1}} \qquad [4.35]$$

According to Fermi's golden rule [GÉR 98], the energy radiated by that molecule at a frequency ω_{EL} (a wavelength λ_{EL}) is proportional to the electromagnetic density of states (DOS) in free space given by:

$$\rho_{rad}(\omega_{EL}) = \frac{1}{3\pi^2}\left(\frac{2\pi}{\lambda_{EL}}\right)^3 \frac{1}{\omega_{EL}} \qquad [4.36]$$

The presence of the NP alters the DOS, and forms another channel for the de-excitation of the molecule. As the NP supports plasmon modes over a small effective volume, the DOS increases in relation to that of free space, by what is known as the Purcell factor [PUR 46]. The DOS for the lth mode at a distance d from the NP is given, in this case, by [KHU 09]:

$$\rho_l(\omega, d) = \frac{L_l(\omega)}{V_{eff,l}}\left(1 + \frac{d}{a}\right)^{-2l-4} \qquad [4.37]$$

where $L_l(\omega)$ and $V_{eff,l}$ are the normalized Lorentzian distribution and the effective volume of the lth mode, respectively.

The Purcell factor, therefore, is given by the ratio between the two DOS:

$$F_{p,l}(\omega) = \frac{\rho_l(\omega, d)}{\rho_{rad}(\omega)} = \frac{3\pi\varepsilon_D(l+1)^2\,\omega L_l(\omega)}{4\chi^3}\left(\frac{\chi}{\chi + \chi_d}\right)^{2l+4} \qquad [4.38]$$

The phenomenon of electroluminescence enhancement by the presence of the NP can be described as a two-step process, as illustrated by Figure 4.12. Firstly, the electrically excited molecule is de-excited with a radiative decay rate τ_{rad}^{-1}, increased by the Purcell factor $F_{p,l}$. This de-excitation, as pointed out previously, is accompanied by a nonradiative decay rate τ_{nrad}^{-1} intrinsic to the molecule. The second step is the coupling of the plasmon modes of the NP with the propagative mode of the molecule. Note that only the 1st-order mode is outcoupled from the NP radiatively, with a coupling efficiency η_{dp}; the higher-order modes ($l > 1$) are nonradiative. Based on this description, the radiative efficiency, or quantum yield, of the molecule in the presence of the NP can be written thus:

$$\eta_{sp} = \frac{\tau_{rad}^{-1} + F_{p,1}\tau_{rad}^{-1}\eta_{dp}}{\tau_{nrad}^{-1} + \tau_{rad}^{-1} + \sum_{l=1}^{\infty} F_{p,l}\tau_{rad}^{-1}}$$

[4.39]

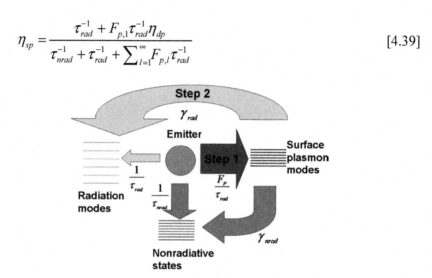

Figure 4.12. *Diagram illustrative the phenomenon of electroluminescence enhancement by the plasmonic effect. From [SUN 08]*

The enhancement factor defined by the ratio between the radiative efficiency in the presence of the NP and that in the absence of the NP is equal to:

$$F = \frac{\eta_{sp}}{\eta_{rad}} = \frac{1 + F_{p,1}\eta_{dp}}{1 + F_{p,1}\eta_{rad}}$$

[4.40]

The first conclusion that we can draw from this description is that the coupling efficiency η_{dp} of the dipolar mode must be greater than the original radiative efficiency of the molecule in order to be able to enhance its luminance. In other words, electroluminescence enhancement is only possible for molecules whose quantum yield is poor to begin with.

Figure 4.13 shows the dependency of that enhancement factor on the radius of the NP and on its distance from the molecule having a quantum yield of 0.01, placed in the vicinity of a gold NP. The enhancement factor depends simultaneously on the size of the NP and its distance from the molecule. The optimum size for the NP is fairly small, in order to have powerful resonance and therefore a low effective volume, but reasonably large to favor radiative outcoupling of the dipolar mode (antenna effect). Similarly, the enhancement factor reaches a maximum for an optimal separation distance d, which is not too great, so as to take advantage of the Purcell factor of the dipolar mode, but not too small either, to prevent coupling of the molecule's energy with the higher resonance modes of the NP (absorption losses).

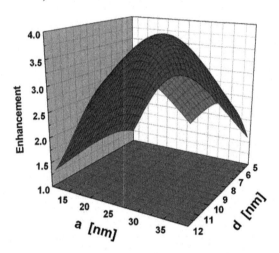

Figure 4.13. *Electroluminescence enhancement factor as a function of the radius of the NP and its distance from the molecule. From [SUN 11a]*

4.6.3. Modification of photoluminescence

Two enhancement mechanisms must be considered in the case of photoluminescence (PL): enhancement of the molecule's absorption and enhancement of its emission. In the presence of a metallic NP, the energy from the source can excite the plasmon modes of the NP, which amplifies the energy in the vicinity of the molecule and thus enhances its absorption. These plasmon modes can also enhance the molecule's emission by means of the Purcell effect. It is important to note that this phenomenon depends heavily on the excitation and emission frequencies of the molecule, and on the plasmon resonance frequency. In other words, PL is greatest when the plasmon resonance frequency ω is situated between the frequency of the excitation source ω_{ex} and that of the emission ω_{PL}. PL enhancement can be achieved by combining the two mechanisms discussed above, and can be explained as follows: the light from the source reaching the molecule–NP system excites both the molecule and the plasmon modes of the NP. A portion of the energy of the dipolar mode is absorbed by the molecule, with an absorption rate γ_{abs}. Next, the molecule undergoes de-excitation with an original radiative decay rate of $1/\tau_{rad}$, which is increased by the Purcell factor F_p due to the presence of the NP. This emission is joined by a portion of the energy of the dipolar mode (the only radiative mode), with a coupling efficiency of η_{dp}. The PL enhancement factor, therefore, is given by the product of the absorption enhancement $F_A(\omega_{ex})$ and emission enhancement $F_L(\omega_{PL})$ factors, as follows [SUN 09]:

$$F_{PL}\left(\omega_{ex},\omega_{ex}\right)=F_A(\omega_{ex})F_L(\omega_{PL})$$
[4.41]

As indicated in the previous sections, the size of the NP is a determining factor in the overall balance. A compromise must be found between the antenna effect, favored in the case of a large dipole (fairly significant size of the NP) and the resonator effect, i.e. good confinement of energy, favored in the case of a small NP. The higher orders of the plasmon modes also have a negative impact on the phenomenon of enhancement. This can be minimized by optimizing the distance between the molecule and the NP.

The PL enhancement factor has been evaluated in the conditions which encourage enhancement of absorption and luminescence – i.e. for the case of molecules with a small effective absorption cross-section and a low radiative efficiency, considering $\omega_{ex}, \omega_{ex} \sim \omega$. It can be given by:

$$F_{PL}=2Q_1^2(1+2Q_1^2)\approx 42Q_1^4$$
[4.42]

where $Q_1 = \dfrac{\omega_1}{\gamma_1} = \dfrac{\omega_1}{\gamma_{rad}+\gamma}$.

Figure 4.14 shows the dependence of that factor on the radius of a nano-sphere of gold. There is a maximum point at the optimal value of the NP size. In addition, the important point to note is that the contribution of the two enhancement factors (absorption and PL) is comparable, which gives a fairly significant overall enhancement factor.

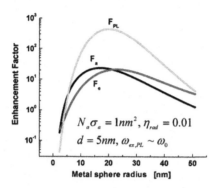

Figure 4.14. *Photoluminescence enhancement factor as a function of NP radius. From [SUN 09]*

4.6.4. *Amplification versus loss: analysis and discussion*

Although many studies have shown that nanoparticles and plasmonic antennas can produce a significant amplification of the PL of adjacent emitters, the mechanisms of enhancement have not yet been fully investigated, and are therefore not yet fully understood. One of the major problems with metals is the losses caused by energy dissipation through the various collisions. Consequently, a number of questions have recently been raised about plasmon amplification and loss compensation. Many projects have shown that the ability of LSPs to improve the fluorescence of emitters placed in proximity to an NP depends heavily on the emitter's quantum yield and its position in relation to the surface of the metal. In particular, the group led by Khurgin [KHU 09, SUN 08, SUN 11a, SUN 09, KHU 12], in their model, described earlier, defend the idea that a high enhancement factor is only possible for emitters whose original quantum yield is poor. They hold that losses in the metal are intrinsic to the metal itself, and are of the same order of magnitude as the damping of the metal. Indeed, the losses cannot be compensated by varying the external factors such as the geometric parameters of the metallic NPs and their environment.

Conversely, though, many experimental studies have demonstrated enhancement of the emission of originally efficient emitters beyond the limit imposed by losses in the metal. For example, G. Lozano *et al.* [LOZ 13] reported substantial improvements of emitters with quantum yields close to 1, using lattices of Al nanoparticles. The authors stress that the total enhancement is due simultaneously to the increase of the dye's absorption efficiency at the excitation frequency when the system is optically pumped, and to the amplification of its emission by modification of the DOS by the plasmonic modes. Improved spatial coherency and highly directional emission have also been observed. This contradicts the idea of the limitation of plasmonic amplification, as mentioned above. Furthermore, in the project conducted by S. Kéna-Cohen *et al.* [KEN 11], it was shown to be possible to reduce the density of triplet states, which are nonradiative and have long lifetimes, whilst preserving the proportion of singlet states, which are less affected. Kéna-Cohen's team used a chain of spherical NPs composed of a silica core coated in a gold shell. Thus, these NPs can be used to inhibit undesirable transitions and reduce the lifetime of the triplet states, which can produce an increase in the repeat rate of pumping of the organic laser.

Note that the aforementioned projects were conducted using optically pumped molecules. Therefore, the enhancement factor is quite high, unlike in the case of electroluminescence, where the excitation is delivered electrically.

4.7. Effect of LSP on the properties of organic sources: state of the art

The use of the plasmonic effect of metallic NPs is one of the most promising ways in which to improve the optical and electrical performances of OLEDs. The choice of metal to use, the size and the position of the NPs are linked to the physical properties the device is intended to have, but also to the approaches and techniques used for their fabrication. Three approaches can be used: Random Metallic Nanoparticles (RMN), Periodic Metallic Nanoparticles (PMN) and Functionalized Metallic Nanoparticles (FMN). In this book, we shall only present RMN and PMN. The details of the FMN approach could be given in a future volume.

The technique of thermal evaporation is widely used in the manufacture of RMNs, given the technological advantages linked to the ease of its use and its suitability for manufacturing large-surface devices. Using this technique, NPs can be deposited in any position within the OLED heterostructure, and we can control their distance from the emissive layer. The size of the NPs can be controlled by varying the thickness of the deposited layer, which builds up in the form of aggregates for thicknesses of less than a few nanometers. On the other hand, this technique does not allow us to control the periodicity and size distribution of the nano-objects. However, the random distribution of the NPs can be exploited to amplify the optical properties of the entities in close proximity to the NPs. Indeed, small NPs act as good resonators, making for good confinement of the EM field, and large ones act as antennas, which outcouple the plasmon modes from the NPs. This situation can lead to the formation of so-called "hot-spots" between the NPs [SUN 11b].

Because of the advantages mentioned above, the existing body of literature includes numerous recent works on improving the yield of OLEDs by using RMN structures. Yang *et al.* [YAN 09] proposed using a cathode incorporating NPs of silver (LiF/Ag clusters/LiF), made with the thermal evaporation technique, to solve the problem of injecting charges between the cathode and the organic layers. These authors showed a 1.75-times enhancement of the photoluminescence of the Alq3 layer. In another publication, the same authors studied the effect of NPs on the phenomenon of Förster energy transfer in an OLED whose emissive layer is a guest–host system of Alq3 and DCM [YAN 92]. They achieved an enhancement of the OLED's emission by a factor of 3.5 using NPs of average diameter 10 nm, exhibiting a LSPR of around 492 nm. This result is attributed, by the authors, to the improvement of Förster energy transfer between the molecules of Alq3 and those of DCM. Another study was conducted by Fujiki *et al.* [FUJ 10], this time inserting NPs of gold onto the ITO anode. The NPs were chemically created by electrostatic absorption, and present an LSPR of around 525 nm. In this study, the emphasis was placed on the importance of the size of the NPs, and of their distance from the emissive layer. An enhancement of the OLED's electroluminescence, and an increase in its internal quantum yield by a factor of 20, were demonstrated. In addition, many works have reported on the study of the position of RMN nanostructures in the organic heterostructure. For example, Xiao *et al.* [XIA 12] demonstrated a 25% enhancement of the electroluminescence of OLEDs incorporating gold NPs (average diameter of 40 nm) in the hole transport layer. These OLEDs, as emission layers, used Alq3 and Alq3:DCJTB, which respectively emit at 525 nm and 600 nm. An increase in current density and a decreased lifetime of the excitons were also demonstrated. Yan *et al.* [YAN 13] studied a tandem OLED emitting in the green spectrum (Alq3 as an emissive layer), incorporating silver NPs in the electron transport layer (Bphen), fabricated by thermal evaporation. The average diameter of the NPs is 20 nm, with an LSPR of around 525 nm. The authors reported an improvement in charge generation, leading to an increase in current density by a factor of 2, and a decrease in the working voltage of the OLED with NPs. A decrease in the lifetime of the excitons was also shown.

PMN-type NPs, for their part, can be produced electrochemically, by nanosphere lithography, nanoprinting or electron-beam lithography. These techniques differ in terms of ease of manufacture, surface area and production time and quality of the NPs. In the case of plasmonic OLEDs, this type of structure exhibits a disadvantage, pertaining to the control of its position inside the organic heterostructure (the OLED), because the PMN structure must first be made on conductive, solvent-resistant substrates. However, there have been many studies on PMNs, showing the effect of the localized surface plasmon on the optical properties of organic emitters. Owing to the technological difficulties, the studies thus far have focused, largely, on optical pumping. Huang *et al.* [HUA 13] reported the enhancement of the photoluminescence of molecules of DCJTB (a red dye) using a lattice of silver nanoparticles. These triangular-shaped nanoparticles were created by nanosphere lithography. A 9.4-fold enhancement of the photoluminescence and a 0.3 ns decrease in lifetime of DCJTB due to the LSPRs of the NPs were shown to take place. Similarly, Bakker *et al.* [BAK 08] studied the properties of a fluorescent dye (Rhodamine 800) in the presence of a periodic dimer lattice of elliptical gold NPs made by electron-beam lithography. An improvement both of absorption and of the emission of the dye in the presence of the NPs was observed. In addition, when the LSPR is at the emission wavelength, the rates of radiative and nonradiative emission change. Furthermore, an alteration of the emission as a function of the detection angle in the presence of the NPs was reported. Also, the LSPR effect of PMN structures has been exploited to improve the spontaneous emission, the coherence and the directivity of organic emitters. Lozano *et al.* [LOZ 13] examined the effect of a lattice of aluminum nanocylinders (with a 400 nm period and made by nanoprinting lithography) on the properties of a dye with a quantum yield near to 1. They were able to obtain a 60-fold enhancement of the photoluminescence at a certain angle of observation. This improvement is mainly attributed to the effect of collective resonance generated by the lattice (known as PLR, for Plasmon Lattice Resonance). In addition, a laser effect with optical pumping was demonstrated using a periodic lattice of plasmonic NPs of Au and Ag, surrounded by a gain medium [ZHO 13, DRI 15]. A lowering of the lasing threshold and a spectral FWHM of < 1.3 nm were obtained. It has been reported that the NP lattice must not only exhibit LSP resonances on the individual NPs, but also satisfy the condition of Bragg resonance to engender a PLR effect.

Besides this, there have been few published studies on PMN structures to improve the performances of OLEDs. K.H. Cho *et al.* [CHO 12] examined PMN structures of Ag in a polymer light-emitting diode (PLED). Hexagonal lattices of cylindrical plots of Ag with a period of 400 nm and 500 nm were nanoprinted on an ITO substrate. This type of lattice facilitates the emergence of the LSPR and PLR modes. The photoluminescence and electroluminescence of the devices made on these structures were enhanced by 70% and 50%, respectively. This was attributed to the improvement of light extraction by Bragg scattering. In a different study, NPs of Ag deposited through a nanoporous film of alumina onto a substrate of ITO were used to improve charge injection [JUN 14]. They showed an improvement of the current density by a factor of 4 and a decrease in the working voltage of the OLED including NPs. This was attributed to the increase in injection of holes by the anode. Furthermore, the intensity of the electroluminescence of the OLED containing NPs was improved by coupling between the surface plasmon of the metallic NPs and the excitons in the emissive layer.

This state-of-the-art shows that the LSPR of RMN or PMN NPs can lead to overall improved performances of the OLED, by improving the injection and transport of charges, the energy transfer in guest–host systems and decreasing the lifetime of the excitons, leading to an increase in the device's electroluminescence and efficiency. The use of PMN-type NPs presents highly interesting optical properties, which can be exploited to improve the directivity and coherence of the emission of OLEDs.

Also, the use of an emitter–metallic NPs system in a microcavity may produce a further improvement of emission, if the wavelength of the LSPR and the mode of cavity coincide. In this case, to our knowledge, there have been very few studies combining the plasmonic effect and the effect of a half-cavity. We note the work of Zhang *et al.* [ZHA 14] and Khadir *et al.* [KHA 15], who studied a plasmonic OLED in a half-cavity using RMN NPs deposited by thermal evaporation onto ITO, which was, itself, deposited on a Distributed Bragg Reflector (DBR). The microcavity is formed of the DBR/ITO/Au-NPs and the OLED's cathode of Al. Their results show that the yield of the OLED deposited on a DBR increases in comparison to that deposited on ITO. In addition, the authors showed a high purity of color of the OLED, attributed to a combination of the effect of the DBR and the LSPR effect of the NPs.

4.7.1. *Study of the effect of random metallic nanoparticles (RMN) on the properties of OLEDs*

In this section, we present a typical example of a recent study on the influence of LSP on the optical and electrical performances of OLEDs. This study is drawn from the work of the PON team at the LPL (Paris 13 University).

The structure of the OLED produced and studied is shown in Figure 4.15. The different layers were deposited, as indicated in Chapter 2, by thermal evaporation on a substrate of glass/ITO. The organic materials deposited are: 4,4', 4", tris- (3-methylphenylphenylamino) triphenylamine (m-MTDATA) as a hole injection layer (HIL) with a thickness of 30 nm, 15 nm of N, N'-diphenyl-N, N'-bis (1-naphtyl)-1,1'-biphenyl-4,4-DiAmi (NPD) as a hole transport layer (HTL), 4-(dicyanomethylene)-2-methyl-6- (p-dimethylaminostyryl)-4H-pyrane (DCM) dispersed at a rate of 1.5% in a host matrix of tris-aluminum (8-hydroxyquinolinato) (Alq3), having a thickness of 30 nm as an emissive layer (EML), exhibiting an emission peak centered at 620 nm. This is followed by 5 nm of Bathocuproine (BCP) as a hole blocking layer (HBL) and 30 nm of 4,7-diphenyl-1,10-phenanthroline (BPhen) as an electron transport layer (ETL). The structure is finished by the deposition of a thin layer of LiF (2 nm) to facilitate the injection of the electrons and a layer of aluminum (Al) 120 nm thick as the cathode.

For the OLED incorporating NPs, a very thin layer of Ag, which forms separate aggregates in the organic heterostructure, was evaporated. The deposition rates for the organic materials, the metal cathode and the layer of Ag are 0.2, 1 and 0.15 nm/s, respectively.

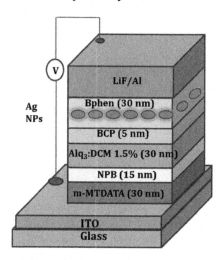

Figure 4.15. *Diagram of the OLED structure incorporating Ag NPs [KHA 15]*

Luminous efficiency is an important value to determine in order to concretely evaluate the effect of NPs on the yield of an OLED. Figure 4.16 plots the luminous efficiency curves as a function of the current density for the four devices under discussion here. We can see that the luminous efficiency of the OLED with 1 nm of Ag is increased by around 17% in comparison to the reference OLED. On the other hand, although the luminance of the device incorporating a 6 nm layer of Ag has improved, its efficiency has fallen considerably. In the latter device, it is not possible to have LSPR–exciton coupling because, as previously mentioned, silver forms a continuous layer which can generate a delocalized surface plasmon but not a localized one. Consequently, dissipation by the metal is the dominant effect in this case.

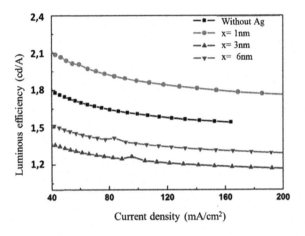

Figure 4.16. *Luminous efficiency as a function of current density for OLEDs without Ag nanoparticles, and with Ag NPs of different thicknesses (x) [KHA 15]*

To confirm and define the correlation between the enhancement of the luminous efficiency and the plasmonic resonance of NPs in the case of the OLED with 1 nm of Ag, the absorption spectrum of the NPs incorporated into the organic heterostructure was compared to the absorption spectrum of DCM and the photoluminescence spectra of Alq3 and DCM. The results obtained are plotted in Figure 4.17. The absorption spectrum of the Ag NPs is fairly broad, comparable to that obtained by simulation. We can see that it shows a central peak at 500 nm and a slight shoulder at 600 nm. This shouldering is in close agreement with the emission peak of DCM. It can lead to an enhancement of the emission of the DCM molecules by coupling between the excitons and the LSPR of the Ag NPs. In addition, the first peak at 500 nm is situated between the emission spectrum of the donor (Alq3) and the absorption spectrum of the acceptor (DCM), with significant overlap between the three spectra. This overlap shows that Förster energy transfer between the donor material (Alq3) and the acceptor material (DCM) can be improved by the plasmonic effect, as previously demonstrated by [YAN 09]. As indicated in Chapter 2 of this book, the rate of Förster energy transfer is expressed as a function of the lifetime of the donor τ_D, the distance between the molecules of the donor and those of the acceptor R, and the Förster radius R_0 as follows:

$$\Gamma_{F\ddot{o}rster} = \frac{1}{\tau_D}\frac{1}{R^6}\left(\frac{3}{4\pi}\int\frac{c^4}{\omega^4 n^4}F_D(\omega)\sigma_A(\omega)d\omega\right) = \frac{1}{\tau_D}\left(\frac{R_0}{R}\right)^6 \qquad [4.43]$$

$F_D(\omega)$ and $\sigma_A(\omega)$ are the normalized emission spectrum of the donor and the effective absorption cross-section of the acceptor, respectively. Indeed, these two values, which determine the Förster radius, can be increased by the LSPR effect of the Ag NPs, provided the three spectra overlap. This leads to an increase in the Förster radius, improving the energy transfer between the donor (Alq3) and the acceptor (DCM).

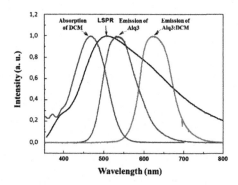

Figure 4.17. *Absorption spectrum of NPs created by depositing 1nm of Ag within the organic heterostructure (black curve). For comparison, the figure also shows the absorption spectra of DCM (blue curve), and the photoluminescence of Alq3 (green curve) and Alq3:DCM (red curve) [KHA 15]. For a color version of this figure, see www.iste.co.uk/boudrioua/lasers.zip*

Finally, the morphological characterization of the NPs of Ag inserted into the OLED using a scanning electron microscope (SEM) (Figure 4.18(a)) shows the formation of clusters of different sizes, randomly distributed across the surface. The size distribution is presented in Figure 4.18(b), and ranges between 8 nm and 20 nm, with an average size of around 10 nm. The mean eccentricity of the NPs is 0.5, as shown in Figure 4.18(c). These observations seem to accord closely with the results, reported above.

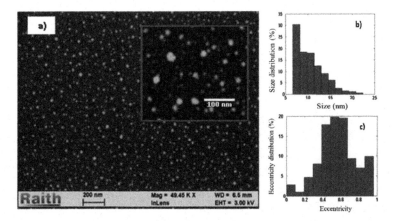

Figure 4.18. *a) SEM image of 1nm thick Ag layer deposited on a Si substrate. b) Size and c) eccentricity distribution of the Ag-NPs [KHA 15]*

4.7.2. *Study of the effect of periodic metallic nanoparticles (PMNs) on the properties of OLEDs*

The potential of Periodic Metallic Nanoparticles (PMNs) to be integrated into various applications has been demonstrated – particularly in the making of new miniaturized optical components. The position and spectral width of the plasmon resonance are largely dependent on the parameters of the NPs, but also on their periodicity in the surrounding medium.

These periodic structures can be divided into two categories, depending on the period of the lattice and the type of plasmon generated: lattices with a short period give rise to near-field coupling, and those with a long period allow for far-field coupling.

For lattice periods smaller than the resonance wavelength, only localized surface plasmon resonances (LSPRs) at the NPs can occur. These resonance modes are due to the plasmon resonances on the individual NPs. If the period of the lattice is small enough, then the NPs are very close together, which can cause an intense EM field in the gap between the NPs. These places are known as hot spots, inducing strong optical near-field coupling [ROS 16]. The presence of emitters in the immediate vicinity of this type of structures can yield an enhancement of their emission by several orders of magnitude. When the period of the lattice becomes comparable to or greater than the resonance wavelength, a second type of plasmon may occur, called plasmon lattice resonance (PLR). The constant of the lattice in this case must satisfy the Bragg conditions. These modes are hybrid plasmonic/photonic in nature, due to coupling between the photonic modes of the crystal created by the periodic arrangement of the NPs and the plasmons generated at each NP. The effect of periodicity is somewhat similar to the case of a photonic crystal, but with greater polarizability at the metallic NPs, inducing strong far-field coupling, which is different to the coupling generated by dielectric NPs [HUM 16, HUM 14]. This type of structure could hold a great deal of interest for enhancing the emitter's spontaneous emission and improving the emission's directivity and coherence.

In the following section, we have opted to present an example of an experimental study of short-period lattices of Al nanorods, made by electron-beam lithography, the principle of which is described below.

4.7.2.1. *Technique of electron-beam lithography*

Electron-beam lithography consists of irradiating a surface covered by an electron-sensitive resin using a focused beam of electrons. The absorption of energy in specific places causes intramolecular phenomena, such as the breaking of the molecular bonds defining the characteristics of the polymer (resin). Electron-beam lithography offers better resolution and precision in comparison to other types of lithography, such as optical lithography, where the resolution is limited by the wavelength of the light and the phenomena of optical diffraction [WOL 04]. A very small resolution of less than 10 nm can be obtained thanks to the finesse of the electron beam. Electron-beam lithography devices are very similar to electron microscopes, and particularly to scanning electron microscopes (SEMs): the principle is to produce a beam of electrons and guide it over the surface of the sample. In reality, numerous lithography devices are converted SEMs, in which a computer controls the movement of the electron beam.

The patterns to be created are designed previously, along with the desired dimensions. Next, the beam irradiates the surfaces occupied by the patterns. In the case that we use a positive resin such as PMMA (Poly(methyl methacrylate)), under the influence of the electron beam, the polymer chain is divided into small molecular units. Using an appropriate development solution, the fractured polymer chains are selectively dissolved. Non-exposed areas remain insoluble in the revealing solution. Consequently, a mask is formed on the substrate. A metallic deposit on that substrate, followed by the lift-off of unwanted parts, gives us plots whose form is determined by the patterns. A recap of the different stages of fabrication of NPs and a photograph of the device used for the task are shown in Figure 4.19.

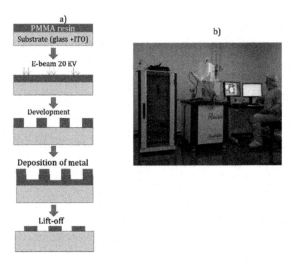

Figure 4.19. *a) Stage of fabrication of metallic nanoparticles by electron-beam lithography. b) Electron-beam lithography system at the Centrale de Proximité en Nanotechnologie de Paris Nord C(PN)2*

4.7.2.2. *Micro-OLEDs: advantage, fabrication and characterization*

Given the fact that electron-beam lithography on large surfaces is very costly in terms of production time, we need to examine the possibility of μ-OLEDs for PMN plasmonic structures. Micro-OLEDs have the advantage of being able to withstand very high current densities, of up to 1 kA/cm² (sometimes even more) with pulsed pumping [NAK 06]. This is greatly advantageous in the search for an electrically pumped organic diode laser, where high current densities are needed to achieve the lasing threshold. In this case, the NPs are fabricated on microstrips of ITO, measuring 1 mm in length and 100 μm in width. These substrates are transferred into the vacuum chamber for deposition of the organic materials and the cathode.

The aluminum cathode is deposited using a mask, which allows aluminum to pass through microstrips of 200 nm in width, which are perpendicular to the ITO strip. The active zone of the μ-OLED is delimited by the intersection between the ITO and Al strips. It has a surface area of 200×100 μm². Four μ-OLEDs can be produced on the same sample (see Figure 4.20). One of these contains the NPs and the others serve as reference OLEDs.

As the active zone of μ-OLEDs is very small, a test setup based on a confocal system is needed to conduct the usual characterizations. The typical diagram and a photograph of this setup are shown in Figure 4.21.

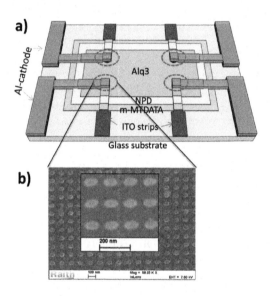

Figure 4.20. *a) Diagram of the final device and b) SEM image of the metallic structures used (example of Al nanorods) [GAR 16]*

The light emitted by the OLED is collected by a microscope objective (20× magnification) and then coupled with an optical fiber using a second objective with lesser magnification (10×), before being focused onto an avalanche photodiode to measure the optical power and/or an optical spectrometer (Ocean Optics USB 2000®) to obtain the electroluminescence spectrum of the μ-OLEDs. The sample is aligned using a halogen lamp in reflection mode, and the image is observed using a CCD camera.

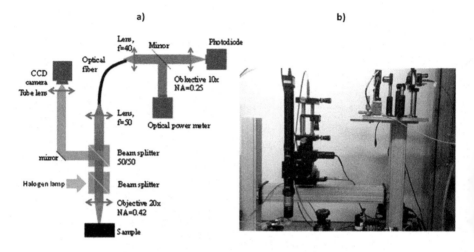

Figure 4.21. *a) Diagram and b) image of the optical setup used for characterizing μ-OLEDs*

4.7.2.3. *Study of a green μ-OLED incorporating aluminum PMN structures*

Green OLED heterostructures were the first OLEDs to be developed 30 years ago. Indeed, this type of heterostructure and the physical phenomena occurring therein are well known in comparison to other types of OLED. This is due mainly to the stability of green emitters such as Alq3.

The architecture of the μ-OLEDs being studied here is represented in Figure 4.22. The organic layers deposited are: 4,4', 4", tris-(3-methylphenylphenylamino) triphenylamine (m-MTDATA) as an HIL, with a thickness of 10 nm, 5 nm of N, N'-diphenyl-N, N'-bis (1-naphtyl) -1,1-diphenyl-4,4-DiAmi (NPD) as an HTL, 80 nm of tris (8-hydroxyquinolinato) aluminum (Alq3) as electron emission- and transport layers, with an emission peak centered at 520 nm. The organic layers are completed by an Al cathode 150 nm thick. In this device, the electron–hole recombination zone is near to the NPD–Alq3 interface, which is situated between the NPs (15 nm from the ITO substrate).

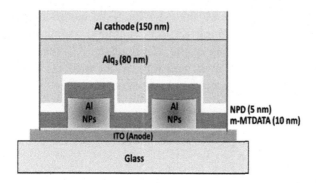

Figure 4.22. *Diagram of the structure of green*
µ-OLEDs incorporating Al NPs [GAR 16]

The characteristics of current density and luminance as a function of the voltage of µ-OLEDs with and without Al NPs are presented in Figure 4.23.

We can see that the current density of the µ-OLED including Al NPs is greater than that of the reference device. The threshold of current density and the working voltage are also reduced in the presence of NPs. This suggests that the incorporation of Al NPs into the µ-OLED affects the charge-transport mechanism, which can be attributed to local enhancement of the electrical field around the NPs, as a result of the LSPR effect [LIU 12, LIU 09]. Also, this improvement of the electrical characteristics of the plasmonic µ-OLED can be attributed to the increase in the active surface area, due to the fact that the deposition of the organic materials and the cathode obey the geometry of the structure of Al NPs, in comparison to the reference µ-OLED, which have flat surfaces.

Furthermore, the variation of the luminance as a function of the voltage exhibits a trend similar to that of the current density. For the same level of voltage, the luminance of the plasmonic µ-OLED is twice that of the reference device.

To demonstrate the effect of NPs on the yield of µ-OLEDs, Figure 4.24 shows the relation between luminance and current density. A 20% enhancement of the efficiency of the µ-OLED incorporating NPs was obtained for a fixed level of current density (the maximum value of the reference OLED). The improvement of efficiency indicates that the luminescence of the Alq3 molecules and their quantum yield are enhanced by the LSPR effect generated by the Al NPs. Two complementary processes may be behind this amplification: namely, spontaneous emission enhancement due to coupling of the LSPR of the Al NPs with the excitons generated in the emissive layer, and improvement of the luminance due to the increase of charge transport, as mentioned previously.

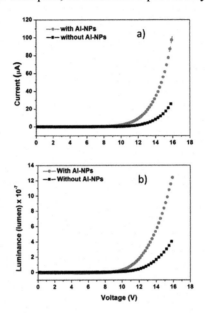

Figure 4.23. *a) Current density and b) luminance as a function of the voltage for green μ-OLEDs, with and without Al NPs [GAR 16]*

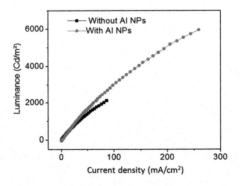

Figure 4.24. *Luminance as a function of the current density for green μ-OLEDs with and without NPs of Al [GAR 16]*

4.7.3. *Study of a plasmonic OLED in a vertical half-cavity*

It is well known that in a cavity, the Purcell effect can reduce the lifetime of the excitons [GU 13], leading to an increase in the rate of spontaneous emission and the quantum yield of the OLED. In addition, the role of the cavity is to increase the directivity of the photons in its resonant mode. Various works have been published on OLEDs in a Fabry–Pérot microcavity, with the aim of optimizing these devices in terms of quality factor, and their compatibility with electrical pumping [MAS 99, BUL 98, DOD 96]. The question which remains pertains to the use of a plasmonic OLED combined with a microcavity.

In this context, we present the results of the study recently conducted on the association of an OLED incorporating 1 nm of Ag (RMN) in the BPhen layer (BPhen (10 nm) / Ag (1 nm) / BPhen (20 nm)) with a microcavity formed of a Distributed Bragg Reflector at the bottom, and an Al electrode as the top mirror (Figure 4.25) [KHA 15]. The DBR is a multilayer dielectric mirror comprising ten pairs of dielectric layers designed to "quarter-wave" dimensions. These pairs are composed of two dielectric layers – one of titanium oxide (TiO_2), which has a high refractive index, and the other of silicon oxide (SiO_2), which has a low one. The dielectric layers finish with a transparent conductive layer of ITO, playing the role, simultaneously, of a high-index layer and a transparent anode (see Figure 4.25). The mirror's reflection is centered at 630 nm, with a maximum reflectivity greater than 99.5% and a stop-band of 100 nm (Figure 4.26).

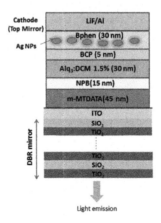

Figure 4.25. *Diagram of the OLED structure in a micro-cavity formed of a bottom DBR and the Al cathode [KHA 15]*

The resonance wavelength of the Fabry–Pérot cavity is determined by the distance between the two reflectors. That being the case, it is determined by the thickness of the OLED and the depths of penetration of the wave into the two mirrors, such that:

$$\frac{m\lambda_{res}}{2} = \sum_i n_i d_i + P_M + P_D \qquad [4.44]$$

where m represents the order of the resonance mode of the cavity, n_i, d_i and $n_i d_i$ are, respectively, the refractive index, the geometric thickness and the optical thickness of each organic layer between the two mirrors. P_M and P_D are the depths of penetration of the wave into the metallic and dielectric mirrors, respectively. It is helpful, of course, to optimize these parameters before carrying out this type of study.

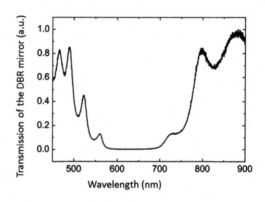

Figure 4.26. *Transmittance of the DBR mirrors used [KHA 15]*

Figure 4.27 shows the electroluminescence (EL) spectra measured through the DBR for the OLED with and without the NPs of Ag. The spectra obtained are also compared to the emission spectrum of a standard OLED.

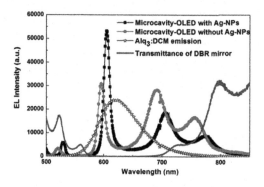

Figure 4.27. *EL spectra of OLEDs with and without Ag NPs in a micro-cavity. The emission spectrum of the OLED without NPs and without a micro-cavity and the transmittance of the mirror are also represented for comparative purposes [KHA 15]*

These results illustrate a significant improvement in the intensity of EL and a reduction in the spectral width of the emission. The EL of the OLED incorporating Ag NPs in a microcavity has improved by a factor of 2 in comparison to the same structure without the NPs. The spectrum obtained presents a peak centered on the resonance wavelength of the cavity with a full width at half maximum (FWHM) of 11 nm, indicating a significant narrowing of the spectrum in comparison to the reference OLED (FWHM = 80 nm). In addition, the spectra of microcavity OLEDs present two additional peaks at 700 nm and 780 nm. These peaks correspond to components of the emission spectrum of the reference OLED, which is fairly broad. The appearance of these peaks is due to the fact that the width of the stop-band of the mirror is less than the width of the EL spectrum of Alq3:DCM.

On analyzing these results, we see that the EL of the microcavity OLED without Ag NPs is 30% greater than that of the OLED without a cavity at the resonance wavelength (605 nm). In addition, the OLED incorporating Ag NPs in a microcavity presents an EL increase of 72% in relation to the microcavity OLED without NPs, and a 123% increase in comparison to the standard OLED.

The 72% increase in luminance is not far off the 87% improvement obtained in the previous section (insertion of 1 nm of Ag into an OLED without a cavity). The significant 123% improvement represents the combined effect of the microcavity and of the plasmonic OLED.

All these results are examples of the work carried out over the past few years on the use of LSPR to improve the optical and electrical properties of OLEDs. Combining plasmonics and organic photonics opens the way to very interesting prospects for the realization of new, high-performance organic sources, and perhaps the possibility of progressing toward an organic diode laser.

One of the nascent ideas – indubitably the most exciting one – is the spaser (*surface plasmon amplification by stimulated emission of radiation*).

4.8. A step toward an organic plasmonic laser?

In this final section, we present a brief introduction to the spaser. Although there are still a number of controversies surrounding this device, it appears to be the ultimate in lasers, capable of generating a coherent plasmonic emission. The principle of operation of this component and the process of its development are discussed below.

4.8.1. *The spaser*

In 2003, Bergman and Stockman [BER 03, STO 09] first introduced the idea of a spaser, which, like a laser, represents a quantum surface plasmon amplifier by stimulated emission of radiation.

This idea involves exploiting a metal/dielectric system to construct a nano-device producing a strong coherent field, confined in a space that is much smaller than the wavelength. In fact, a spaser is the nanoplasmonic counterpart to a laser, but it uses plasmons, and therefore does not emit photons. It is similar to a conventional laser, but in a spaser, the photons are replaced by surface plasmons and the resonant cavity is replaced by a metallic nanoparticle which supports the plasmonic modes.

In fact, the introduction of the concept of a spaser shows how it may be possible to overcome the limit of Rayleigh scattering and confine an EM wave in a space smaller than its own wavelength. The answer lies in the fact that at the nanometric scale, the optical fields are almost purely electrical oscillations at optical frequencies, where the component of the magnetic field is low and does not play a major role in the physical interactions at that scale.

The capacity of a nanomaterial to withstand and concentrate such fields is due to the existence of optical modes, localized at dimensions much smaller than the optical wavelength.

There have been many published works giving further clarifications on the properties of this concept, and studying the device both theoretically and experimentally [STO 08, STO 10, STO 11a, STO 11b, AND 11, LIU 11, NOG 09]. It is interesting to note the following properties:

– unlike a laser, which amplifies photons, a spaser amplifies surface plasmons;

– both laser and spaser systems are based on the principle of stimulated emission and population inversion. However, the spaser generally works in what are known as "dark" optical modes, which cannot be coupled with the far field;

– the dimensions of the cavity of a spaser are mainly determined by the thickness of the metal skin (of the order of a few tens of nm), which explains the sub-wavelength confinement.

In a conventional laser, the emission direction is dictated by an external resonator, and its coherence is linked to the stimulated emission of atoms in the gain medium. In a spaser, the emission direction is normal to the plane of the matrix, where powerful currents trapped in the plasmonic resonators oscillate in phase. The coherence, in this case, results from the fact that the collective oscillations are in phase.

Additionally, the creation of deliberate asymmetry in the plasmonic system enables us to break with the nonradiative nature of dark-mode oscillation, and let a fraction of the energy accumulated in the oscillations radiate out into free space. This is comparable to the leakage from an outcoupler on a laser resonator.

Therefore, unlike with an optical quantum generator, the spaser can be considered a conventional system from all points of view apart from the mechanism responsible for producing the gain, which employs an interaction between the medium and the plasmonic modes [SHE 08].

One of the simplest types of nanoparticles, and potentially one of the most promising for use as a spaser resonator, is a metal–dielectric nano-shell [STO 08]. Such nano-shells have recently been studied, and have a very broad range of applications [HIR 03, AVE 97]. A possible design for a nano-shell spaser is illustrated in Figure 4.28 [STO 08]. It is composed of a nano-shell of silver, surrounded by a number of monolayers of nanocrystals known as nano-quantum dots (NQDs).

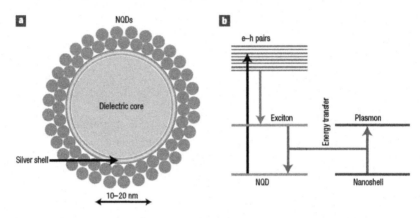

Figure 4.28. *Diagrammatic representation of the operational principle behind a spaser, made of a silver nano-shell over a dielectric core (10-20 nm radius) and surrounded by two dense monolayers of nanocrystalline quantum dots (NQD) [STO 08]. For a color version of this figure, see www.iste.co.uk/boudrioua/lasers.zip*

The external radiation stimulates a transition, resulting in the creation of electron–hole (e–h) pairs (vertical black arrow). The (e–h) pairs sink to lower excitonic levels (green arrow). The exciton recombines, and its energy is transferred (without radiation) to plasmonic excitation of the metallic nanoparticle (nano-shell) by coupled resonant transitions (red arrows) – see Figure 4.28 [STO 08].

In a nano-quantum dot crystal, the excitons would recombine to form photons. However, when the NQD is at the surface of a resonant nanoparticle, the excitonic energy is transferred, with no significant radiation emission, to the resonant LSPs of the nanoparticle (coupled red arrows in Figure 4.28(b)). This process is far more probable, by an entire order of magnitude. The LSPs stimulate other transitions in the gain medium, leading to the excitation of identical LSPs in the same mode, causing the spaser to operate [STO 08].

As previously indicated, if the symmetry of the plasmonic nanoparticle in the spaser is slightly broken, an LSP spasing mode will become radiating. A collection of such spasers, therefore, may begin to emit light, just like a laser. This idea was put forward by N. Zheludev *et al.* [SHE 08], using a meta-material containing a planar lattice of spasers, each with slightly disturbed symmetry. This lattice then becomes a very effective planar laser, emitting light in a direction normal to its plane.

Another example illustrating the operation of a spaser is that reported by Noginov *et al.* [NOG 09]. Nanoparticles 44 nm in diameter, composed of a gold core surrounded by sodium silicate and dye-doped silica shells, are used to make the spaser. In order to create a veritable nanolaser, the authors decoupled the LSPs from photonic modes at a wavelength of 531 nm, making this system the smallest nanolaser reported to date, and the first one to operate at visible wavelengths.

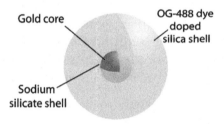

Figure 4.29. *Diagram of the principle of spaser nanoparticles, composed of a gold core, supporting plasmon modes, surrounded by a silica shell containing the organic dye Oregon Green 488 (OG-488), creating gain [NOG 09]*

To study the stimulated emission, samples were loaded into crucibles with a 2 mm-long path, and pumped at the wavelength of 488 nm with 5 ns pulses of a slightly focused optical parametric oscillator. For example, Figure 4.30 shows the appearance of a narrow peak at 531 nm once the pumping energy surpasses the critical threshold value.

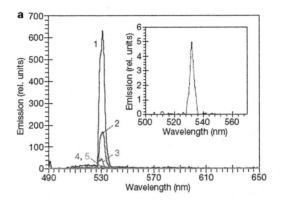

Figure 4.30. *Stimulated emission spectra of spaser nanoparticles pumped at different energy levels: 22.5 mJ (1), 9 mJ (2), 4.5 mJ (3), 2 mJ (4) and 1.25 mJ (5), with a parametric oscillator with 5 ns pulses at 488 nm [NOG 09]*

Figure 4.31 shows the intensity of that peak as a function of the pump energy, giving an input–output curve with the threshold characteristic typical of lasers. Similarly to lasers, the transition of the emission spectrum above the threshold (from a wide band to a narrow line) suggests that the majority of excited molecules are contributing to the stimulated emission. This phenomenon involves resonant energy transfer from the excited molecules to surface plasmon oscillations, and stimulates the emission of surface plasmons in a radiating mode, validating the original concept of the spaser.

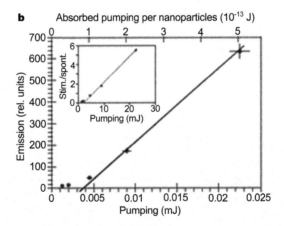

Figure 4.31. *Input–output curve corresponding to a spaser nanoparticle. Boxed insert: ratio of stimulated emission intensity (integrated between 526 nm and 537 nm) to the amount of spontaneous emission (integrated at <526 nm and >537 nm) [NOG 09]*

It is worth noting that this field continues to arouse a great deal of interest and debate. Amongst the theoretical developments of the concept, a nano-lens spaser has been put forward [LI 05], which has a nano-focus ("hottest point") of the local fields. Another nanolaser (spaser) was reported to use nanoparticles of gold and nanorods of InGaN/GaN as a gain medium [WU 11]. This spaser generates emissions at green wavelengths. Also worthy of note is the concept of a distributed feedback (DFB) spaser [MAR 11]. A nanolaser with optical pumping of a semiconductor, coupled with a hybrid/metal semiconductor (CdS/Ag) waveguide, SPP, has been demonstrated with a mode of confinement in transverse mode (in two dimensions), of the order of 10 nm in size [OUL 09].

Obviously, we are left with the question of electrical pumping of a spaser, which echoes the problem posed by electrical pumping of organic semiconductors. This question has also attracted a great deal of interest, and a few works have been published on the subject. For example, we could cite the theoretical work on an electrically pumped spaser (nanolaser) reported in [CHA 08]. Another noteworthy publication is the study of an electrically pumped nanolaser with a semiconductor gain medium [HIL 09], where the laser modes are SPPs with one-dimensional confinement, of the order of 50nm in size.

An electrically pumped spaser would be particularly useful, and would constitute a revolution in the field of nano-photonics [FED 12]. Indeed, actively generating coherent surface plasmons could yield new opportunities for the manufacture of photonic metamaterials with an indubitable impact on the technological developments to exploit the optical and plasmonic effects at the nanometric scale.

4.9. Conclusion

This chapter has been devoted to the study of the plasmonic effect – in particular, the phenomenon of the localized surface plasmon (LSP) generated by metallic nanoparticles. To begin with, we briefly described the response of metals subjected to an EM field, considering the different models describing the dielectric functions. We also demonstrated that the size, shape, nature and environment of metallic NPs have significant effects on their optical properties. Two types of nanoparticles were considered – RMNs and PMNs. In the first part of the chapter, the effects of RMNs of silver, made by thermal evaporation and inserted into OLED heterostructures, were presented. For the optimized parameters, a 20% enhancement of the efficiency of OLEDs including NPs was shown. In the second part, periodic lattices of metallic NPs of Al, made by electron-beam lithography on glass/ITO substrates, were used with μ-OLEDs. The inclusion of the NPs enhanced the efficiency of the μ-OLEDs by a rate of nearly 20%. Finally, the insertion of a plasmonic OLED into a vertical half-cavity enhanced the electroluminescence by a factor of 2, accompanied by a spectral narrowing to 11 nm.

In the final section, we succinctly presented the operational principle behind a spaser (surface plasmon amplification by stimulated emission of radiation). The making of this nanolaser will have an undeniable impact on the technological developments of organic nano-photonics.

Conclusion

Given the recent progress made in organic optoelectronics, we can expect promising development in this area. The emerging organic optoelectronic components arousing increasing levels of interest from industrialists are, at present:

– organic photovoltaic (OPV) plastic cells;

– organic light-emitting diodes (OLEDs);

– organic field-effect transistors (OFETs) in thin layers;

– organic laser diodes (OLDs).

One of the biggest issues in organic photonics is the realization of the first organic laser diode (electrically pumped), which has enormous potential for numerous applications (compact, low-cost tools for spectroscopy, chemical sensors, micro-fluidics, etc.). These lasers can, indeed, be tuned to emit over a truly vast spectral range – in particular, covering the whole of the visible spectrum. They can be manufactured very easily and cheaply, and can be deposited in the form of thin films on any type of substrate (flexible or rigid).

Numerous organic lasers have been made with optical excitation. The results obtained have facilitated remarkable progress toward reducing the lasing threshold and demonstrating the potentials of these organic lasers. However, various physical processes (low charge mobility, absorption, guiding, etc.) have hitherto prevented electrical pumping from becoming a reality.

To overcome these obstacles, particular attention is paid to the optical aspect, placing the OLED in a microcavity with a high quality factor, thus reducing the losses linked to poor confinement of the optical field inside the device. However, these studies involve the filtering of the OLED's emission spectrum by the microcavity. According to this principle, all the photons emitted which are not coupled with the microcavity's resonance mode are lost, produced by charges which needlessly increase the lasing threshold. In this context, to increase the coupling rate it is important to increase the OLED's optical gain to the resonance wavelength of the microcavity, in order to facilitate a low-threshold approach to organic laser diodes.

In order to produce net gain (i.e. gain > losses) and a high current density ($<1kA/cm^2$) with electrical pumping, there are two possible approaches.

The first is to increase the current density achieved by OLEDs, in the heterostructures, which are conventionally made with small organic molecules or using polymer layers offering relatively high mobility. Small molecules (monomers and dimers, etc.), deposited by vacuum evaporation, can be used to create complex heterostructures containing up to ten layers and exhibiting good electron–hole equilibria. However, these materials have the drawback of having low charge carrier mobility, which limits the current density the OLED can achieve. Thin layers based on large molecules, such as polymers, are more conductive, and capable of delivering a relatively high current density; however, they generally cannot be vacuum evaporated with the aim of making thin layers. This limitation is the result of their large molecular weight, combined with their thermal fragility. The process of deposition of these materials consists of finding an appropriate solvent and depositing the polymer solution by centrifugation. The main drawback to this method lies in the difficulty of creating heterostructures, owing to the low number of available multilayer solvents/nonsolvents, at the risk of deteriorating the underlying layers.

The second strategy is to introduce nanoparticles into the OLEDs, exhibiting surface plasmon resonances which enhance the spontaneous emission of the organic compound of the emissive layer. The plasmonic effect seems an effective solution to increase the gain of the organic medium and thereby obtain lower lasing thresholds. The plasmonic effect has recently aroused increasing interest in the field of nanotechnology, with possible applications ranging from detection, biomedicine and imaging to information technology. This increased interest is due, amongst other things, to Noginov *et al.*'s demonstration, in 2009, of a device known as a SPASER (*surface plasmon amplification by stimulated emission of radiation*), capable of generating plasmonic coherent light. It has been proposed that, in the same way that a laser generates a stimulated emission of coherent photons, a SPASER could generate a stimulated emission of surface plasmons in resonant metallic nanostructures surrounded by a gain medium. However, the attempts to make an efficient SPASER are faced mainly with the problems of absorption losses in the metal, which are particular high at optical frequencies. Organic materials seem to be a good candidate to compensate for this loss, and even amplify the surface plasmon, thanks to very high gains of more than 1500 cm^{-1}.

In both cases – deposition of polymer layers or use of nanoparticles – the challenge is to find a technique for deposition of thin layers of polymers or isolated molecules or nanoparticles in a complex OLED heterostructure.

Full knowledge of the photo-physical properties of the organic medium containing the nanoparticles, therefore, would enable us to optimize the nanoparticle–chromophore system. Of these properties, we can cite: the time constants of the different states (the vibrational states, singlet states and triplet states), the gain of the organic medium and its emission spectrum, the effective absorption- and emission cross-sections. In order to study these rapid intra- or intermolecular photo-physical processes induced by the light or more generally when a molecule is excited, such as passages between the energy levels during relaxation and inter-system crossing (between the singlet states and triplet states), the internal conversion of which can reach the scale of a picosecond, it is necessary to use a time-resolved system.

Finally, the quest for an organic laser diode involves numerous laboratories throughout the world, and includes different, pluridisciplinary approaches. This issue is both a complex and exciting one, and we hope this book is a modest contribution to illuminating the readers and providing responses as to the questions and issues in this blossoming field.

Appendix

A Brief History of Organic Lasers

This appendix is taken from the authors' article recently published in the journal Photoniques, "*Les lasers organiques : une quête de 50 ans*", Revue Photoniques 68, 30-33 (2013).

In 1966, a few months apart, P.P Sorokin and J.R. Lankard at the IBM research labs in Yorktown Heights (NY), on the one hand, and F.P. Schäffer, W. Schmidt and J. Wolze at the University of Marburg, on the other, published the first demonstrations of the lasing effect using organic materials in solution as a gain medium. They were followed soon after by B. H. Soffer and B. B. McFarland who, in 1967, presented organic lasers based on rhodamine 6G in solution in ethanol, but also in a solid matrix of polymethyl methacrylate (PMMA). From then on, organic lasers have exhibited line widths of 6 nm (Q = 95) in a cavity with a dielectric mirror, which can reach as low as 60 pm (Q = 9500) if the cavity ends in a lattice.

It is interesting to note that P.P. Sorokin uses chloro-aluminum phthalocyanine in solution in ethyl alcohol, which is an organic material which we find almost identically in OLEDs. The early 1990s saw a renewal of dye lasers with new advances: they no longer operate in a pulsed regime, but also in continuous wave (CW) regime, with lasing thresholds slightly under 550 W/cm^2 and quality factors Q of around 50. Meanwhile, laser diode pumping took the place of arc lamp pumping or pumping by another laser, meaning it became possible to construct more compact lasers.

The greatest innovation at the time, though, from the point of view of the laser diode stemmed from the transparency of polymers such as PMMA, HEMA and ORMOSIL, enabling us to renew and improve organic gain media. Thus, instead of placing the dye in solution, it can be dispersed in a solid host matrix made of one of those polymers, which helps increase the index of the gain medium. It is then possible to construct even more compact lasers and overcome the problems of fluidics, helping reduce the excitation density at the lasing threshold. Although Soffer and McFarland, in 1967, presented organic lasers in a PMMA matrix, Maslyukow *et al.* in 1995 developed solid-state lasers from different rhodamines in a PMMA matrix and obtained laser emissions with FWHM values of 6nm; this represents a quality factor of the order of $Q \sim 100$ with a lasing threshold of only 4 mJ/cm^2 for 10 ns pulses.

Organic gain media in the solid state constitute an important advance for organic lasers, but also as a stepping stone toward the organic laser diode for at least two reasons. First of all, it is possible to obtain the lasing effect in the solid state not only with polymer matrices, but also with small molecules, including pi-conjugated molecules which exhibit semiconductive properties that are interesting for OLEDs, as we shall soon see. In 1997, Kozlov and Forrest's team at Princeton University, in *Nature*, published the proof of laser emission in a vacuum-deposited thin organic layer, composed of DCM2   (4-(Dicyanomethylene)-2-methyl-6-(4-dimethylaminostyryl)-4H-pyran), co-evaporated with Alq3 (Tris-(8-hydroxyquinoline)aluminum). Laser cavities are ribbon-type guiding structures, heterostructures or DFB resonators. These two co-evaporated materials constitute what we call a Guest–Host system. The emission spectrum of Alq3 has a wide overlap with the absorption spectrum of DCM2. This indicates an efficient Förster energy transfer between the Alq3 and DCM2 dipoles. The advantage of the Guest–Host system is to limit absorption losses at the whole system emission. These materials are typically those which are used to make the emissive layers of OLED hereto-structures, as we shall see later on.

In the best configurations, the lasing thresholds can drop to 1 mJ/cm^2. The results published by Forrest's team at Princeton constitute a major advance: indeed, these small organic molecules co-evaporated in a vacuum in thin films or in a polymer matrix deposited by spin coating are conformable and structurable by processes similar to those used in the microelectronics industry, and therefore it is possible to make a wide variety of compact laser cavities on varied substrates. Thus, numerous teams the world over have put forward and studied new structures of organic lasers whose gain medium is organic materials used in OLED heterostructures, and whose cavity can be any of a whole range of cavities even broader than that already considered for inorganic III-V semiconductors.

Given the range of different cavities which have been tested, in order to study their impact on the lasing threshold, we have categorized them using two criteria: the first is the 1D or 2D dimension of the confinement; the second is the axis of the cavity, which may be perpendicular or contained in the plane of the organic thin layer. These different laser cavities are presented in Chapter 3.

Of the 1D confinement cavities and whose axis of light propagation is contained in the plane of the thin layer, besides the ribbon waveguides reported by Kozlov, we must cite the Distributed FeedBack (DFB) cavity introduced in 1971 by Kogelnik and Shank, with rhodamine. Helliotis in 2003, and Riechel in 2000, reported significant advances in terms of reduction of the threshold with this type of cavity. Of the experiments collected in the literature, we can cite those of Tsutsumi and Yamamoto who, in 2006, created a DFB structure using a dynamic holographic lattice, and thus obtained the lasing effect with a qualify factor $Q = 2160$, and a lasing threshold of only 130 $\mu J/cm^2$. Their primary intention was to show that organic gain media using several organic compounds allowing Förster energy transfer lead to a significant reduction in the lasing threshold.

2D DFB structures are capable of further reducing the lasing threshold. In 2000, Riechel et al. at the University of Munich compared 1D and 2D DFBs and showed that the latter have much lower thresholds and greater power conversion efficiency, with a 315 $\mu J/cm^2$ threshold for $Q = 1920$. In 1999, Meir et al. obtained 40 $\mu J/cm^2$ with $Q = 2100$. In 2005, Barnes and his team used a 2D DFB structure, made on a SiO_2 substrate and a lasing threshold of 50 $\mu J/cm^2$, with a quality factor $Q = 855$.

Note that DBR structures obtained by etching gallium nitrate, with a gain medium of coumarin, were tested by Vasdekis in 2006, and they are able to deliver quality factors $Q \sim 1000$ and thresholds of the order of 6 $\mu J/cm^2$: one of the lowest ever thresholds.

One of the lowest lasing thresholds to date is the DFB structure proposed in 2007 by Karnutsch et al., combining two lattices with two different periods, offering two orders of diffraction and reducing the threshold to 36 nJ/cm^2.

Defect cavities in a photonic crystal are more than one variant of 2D structures. Indeed, they offer not only good quality factors but also small modal volumes, which is very favorable for the Purcell effect. They are considered to be favorable for low-threshold laser emission. This type of cavity has been tested as a thresholdless laser in other circumstances. In 2010, Murshidy et al. [MUR 10] report a laser emission from L3 defect in a photonic crystal in a silicon nitride membrane and characterized by a quality factor Q = 1200. In 2005, Kitamura et al. made a photonic crystal in a silicon oxide membrane (Q ~ 1000) and deposited the Alq3:DCM2 host–guest system to obtain a laser emission. In 2012, Gourdon et al. obtained the laser effect with an etched photonic crystal in silicon nitride and covered with a guest–host system of Alq3 and DCJTB (4-(dicyanomethylene)-2-t-butyl-6(1,1,7,7-tetramethyljulolidyl-9-enyl)-4H-pyran), offering a quality factor greater than Q > 350, and with a lasing threshold reduced to only 9.7 μJ/cm^2, which is amongst the lowest thresholds to be obtained, at the time of writing.

Fabry–Pérot microcavities made with multilayer dielectric mirrors, for their part, offer 1D confinement along an axis perpendicular to the plane of the layers, which enable us to achieve high quality factors. Amongst the first results which show a reduction in the threshold, Forrest and his team, in 1998, reported the laser effect in such a microcavity characterized by Q = 420, enabling us to achieve a reduction of threshold excitation density to 300 μJ/cm^2. In 2005, Koschorreck et al. made further progress and demonstrated the lasing effect by placing a gain medium composed of Alq3 and DCM2 between two multilayer dielectric mirrors and thus obtained a quality factor Q = 4500 and a threshold of only 20 μJ/cm^2.

On the basis of this set of experiments, we can highlight a trend which, even approximately, could help find a research strategy to produce an organic laser diode. All these results are shown in Figure A.1, where the horizontal axis shows the quality factor and the vertical axis shows the corresponding lasing thresholds. From this cloud of points, we extrapolate a trend, represented by the boundary between the blue and white zones. This trend is not a law, but reflects the efforts made to drive down the lasing threshold through all these experiments, notably by increasing the quality factor Q. If this trend exists, it shows that when we increase the quality factor, the lasing threshold decreases.

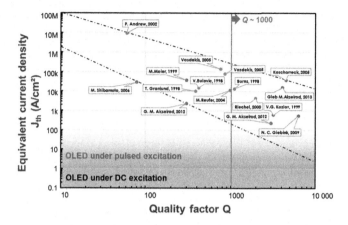

Figure A.1. *Laser threshold as a function of the quality factor for different experiments reported in the literature*

To see the correspondence between the electrical and optical excitation densities, it is necessary to understand the mechanism whereby electrons become photons. We need at least as many excitons electrically provided as there are excitons measured by optical pumping at the threshold. In reality, though, with electrical pumping, the current injected in organic layers brings in charges or excitons with random spin. Yet only excitons with opposing spin are able to recombine, so only $\eta_{quant} = 25\%$ of electrons injected (singlet excitons) lead to radiative recombinations. Thus, there is an initial factor which increases the current density necessary to attain the threshold. In addition, as certain experiments have shown, the external quantum yield varies as a function of different parameters such as the pulse duration, the electrical adaptation of the circuit and of the OLED, and the different materials used to fabricate the OLED heterostructure. An estimation of the current density needed to attain the lasing threshold is:

$$J_{th} = \frac{1}{\eta_{internal}} \frac{q}{h\nu} I_{th}$$

I_{th} is the optical excitation density, q is the charge of the electron, $h\nu$ the energy of a photon, $\eta_{internal}$ is the internal quantum yield calculated on the basis of the external quantum yield, taking account of the degree of coupling, which is the fraction of photons which are truly outcoupled from the component. This coefficient integrates the quantum yield $\eta_{quant} = 25\%$.

In Figure A.1, the excitation densities at the threshold with optical pumping, converted into equivalent current density, are represented for each experiment as a function of the quality factor. The maximum current densities that the OLEDs can withstand in a DC regime and an AC regime are represented by green and yellow rectangles, respectively.

Thus, if laser microcavities with quality factors of the order of $Q = 10,000$ to $Q = 20,000$ were obtained, the expected laser excitation thresholds could be equivalent to the current densities of OLEDs supplied in DC. Thus far, these quality factors have yet to be obtained with electrically pumped organic materials, owing to the metallic electrodes, which add absorption losses. The fact that the current densities it is possible to achieve in DC are at least three orders of magnitude lower than the lowest lasing thresholds is a good indication of the difficulty of achieving the laser effect by DC electrical pumping. On the other hand, in the pulsed (AC) regime, the limit of 800 A/cm^2 is very close to those of the lowest lasing thresholds – in particular, that found by Vasdekis in 2006. An initial objective, therefore, would be to make a microcavity OLED with a quality factor in the range $Q = [4000\text{-}5000]$ and with a current density of around 1–2 kA/cm^2.

To explain the problem of high current densities, it is important, first, to explain how an electrical current can pass through an organic material initially reputed to be an insulator, which leads us to first consider the variety of organic materials which are conductive polymers. This step is important to then go on to touch on organic light-emitting diodes.

It is important to first consider conductive polymers. The work on conductive polymers carried out by the chemists Heeger, MC. Diarmid and Shirakawa in the early 1970s consisted of doping a polyacetylene with an alkali or a halogen. The model polymer is the polyacetylene molecule composed of the carbon-hydrogen pattern [C-H] repeated thousands of times, or even hundreds of thousands, in a long, linear molecular chain. It should be noted that the carbon atoms are connected to one another in an alternating pattern of single and double bonds, which constitutes a conjugated system where the electrical orbitals overlap. This facilitates the delocalization of the electrons on the set of atoms which participate in the overlapping of the orbitals and can therefore facilitate electron transport along the molecular chain that makes up the polymer. However, as it stands, the polymer, in the model of an intrinsic semiconductor, remains an insulator or very poorly conductive because the double conjugated bonds are not sufficient. To become conductive or semiconductive, the polymer must be "disturbed" either by removing electrons (oxidation) or adding them (reduction). This is why, when the polymer is oxidized by a halogen such as iodine, which is the equivalent of P doping, or reduced by an alkali metal such as sodium, which corresponds to N doping, the molecule becomes conductive. The increase in conduction is not linked to the anions of iodine or cations of sodium which move, but to the electrons, which can shift along the conjugated bonds – i.e. along the entire length of the chain. From one chain to another, conduction takes place in jumps. Thus, M.C. Diarmid and Heeger at the University of Pennsylvania and Shirakawa of Kyoto University, in 1977, published a mobility of 3000 $S.m^{-1}$ for iodine-doped polyacetylene. This is an increase of seven orders of magnitude in relation to non-doped polyacetylene. In 2000, all three received the Nobel prize in chemistry for "the discovery and development of conductive polymers".

Small linear molecules, or cyclic conjugated monomers or dimers also exhibit conduction, but lower by the order of 10^{-5} $S.m^{-1}$ because of the increase in the degree of conduction per hop. In thin layers, this mobility is sufficient to facilitate currents of a few tens of milliamperes when subjected to a few tens of volts.

We can distinguish two ways to create thin layers of a few tens of nanometers: the wet and the vacuum evaporation methods.

With the wet technique, we deposit polymers or small light-emitting molecules in solution, by spin coating. The amorphous thin layers of organic materials obtained after evaporation of the solvent are deposited successively on a transparent and conductive electrode. These successive depositions, completed by a metallic electrode, constitute an organic device emitting light when it is traversed by an electrical current. However, this strategy presents significant limitations because the number of successive layers that can be deposited is reduced; this limits the efficiency of the component. Indeed, the solvent of each layer must be an anti-solvent of the previous layers so as not to modify them. Hence, this strategy will not be discussed further, even though it is the subject of a commercial application.

The vacuum evaporation technique, for its part, applies to small molecules which, unlike polymers, it is possible to evaporate. There is no longer any limit on the number of layers deposited. This is why, in spite of their poorer conduction, they are used for the creation of complex organic heterostructures which, ultimately, are characterized by better yields.

We can make a projection that in years to come, the generation of ultra-short electrical pulses (ns and sub-nanoseconds) will enable us to achieve current densities of some 10 kA/cm^2, no longer just in a transport monolayer, but in a high-performing OLED made up, for example, of doped injection and transport layers. In these conditions, OLEDs placed in laser cavities with a quality factor of only $2000 < Q < 3000$ enable us to envisage the electrically pumped laser effect with lasing thresholds of the order of ten kilo-amperes per square centimeter.

Bibliography

[ADA 88] ADACHI C., TOKITO S., TSUTSUI T. *et al.*, "Electroluminescence in organic films with three-layer structure", *Japanese Journal of Applied Physics*, vol. 27, pp. 269–271, 1988.

[ADA 90] ADACHI C., TSUTSUI T., SAITO T., "Confinement of charge carriers and molecular excitons within 5-nm-thick emitter layer in organic electroluminescent devices with a double heterostructure", *Applied Physics Letters*, vol. 57, pp. 531–533, 1990.

[ADA 08] ADAWI A.M., LIDZEY D.G., "Enhancing the radiative decay rate of fluorescent organic molecules using micropillar microcavities and optical nanocavities", *Materials Science and Engineering: B*, vol. 149, pp. 266–269, 2008.

[AKS 14] AKSELROD G.M., YOUNG E.R., STONE K.W. *et al.*, "Reduced lasing threshold from organic dye microcavities", *Physical Review B*, vol. 90, pp. 035209,1–035209,8, 2014.

[ALI 04] ALI T.A., JONES G.W., HOWARD W.E., "Dual doped high Tg white organic light emitting devices on silicon", *Proceedings of the Society for Information Display*, vol. 35, pp. 1012–1015, 2004.

[AND 02] ANDREW P., TURNBULL G.A., SAMUEL D.W. *et al.*, "Photonic band structure and emission characteristics of a metal-backed polymeric distributed feedback laser", *Applied Physics Letters*, vol. 81, pp. 954–956, 2002.

[AON 07] AONUMA M., OYAMADA T., SASABE H. *et al.*, "Material design of hole transport materials capable of thick-film formation in organic light emitting diodes", *Applied Physics Letters*, vol. 90, pp. 183503–183503-3, 2007.

[BAB 03] BABA T., SANO D., "Low-threshold lasing and Purcell effect in microdisk lasers at room temperature", *IEEE Journal of Selected Topics in Quantum Electronics*, vol. 9, pp. 1340–1346, 2003.

[BAL 99] BALDO M., O'BRIEN D., THOMPSON M. *et al.*, "Excitonic singlet-triplet ratio in a semiconducting organic thin film", *Physical Review B*, vol. 60, pp. 14422–14428, 1999.

[BAL 02] BALDO M.A., HOLMES R.J., FORREST S.R., "Prospects for electrically pumped organic lasers", *Physical Review B*, vol. 66, pp. 035321–035337, 2002.

[BAR 75] BARLTROP J.A., COYLE J.D., *Excited States in Organic Chemistry*, John Wiley & Sons Ltd, London, 1975.

[BAU 07] BAUMANN K., STÖFERLE T., MOLL N. *et al.*, "Organic mixed-order photonic crystal lasers with ultrasmall footprint", *Applied Physics Letters*, vol. 91, pp. 171108–171111, 2007.

[BER 97] BERGGREN M., DODABALAPUR A., SLUSHER R.E. *et al.*, "Light amplification in organic thin films using cascade energy transfer", *Nature*, vol. 389, pp. 466–469, 1997.

[BRA 02] BRABEC C.J., SHAHEEN S.E., WINDER C. *et al.*, "Effect of LiF/metal electrodes on the performance of plastic solar cells", *Applied Physics Letters*, vol. 80, pp. 1288–1290, 2002.

[BRO 00] BROWN T.M., FRIEND R.H., MILLARD I.S. *et al.*, "LiF/Al cathodes and the effect of LiF thickness on the device characteristics and built in potential of polymer light emitting diodes", *Applied Physics Letters*, vol. 77, pp. 3096–3098, 2000.

[BRÜ 11] BRÜCKNER R., SUDZIUS M., FRÖB H. *et al.*, "Saturation of laser emission in a small mode volume organic microcavity", *Journal of Applied Physics*, vol. 109, pp. 103116–103116-5, 2011.

[CAL 10] CALZADO E.M., BOJ P.G., GARCÍA D.M.A., "Amplified spontaneous emission properties of semiconducting organic materials", *International Journal of Molecular Sciences*, vol. 11, pp. 2546–2565, 2010.

[CAP 10] CAPELLI R., TOFFANIN S., GENERALI G. *et al.*, "Organic light-emitting transistors with an efficiency that outperforms the equivalent light-emitting diodes", *Nature Materials*, vol. 9, pp. 496–503, 2010.

[CAR 12] CAREY F., *Advanced Organic Chemistry: Part A: Structure and Mechanism*, Springer, 2012.

[CHA 06] CHAKAROUN M., ANTONY R., DEMADRILLE R. *et al.*, "Comparative study of optoelectronic properties of various Europium complexes used in organic electroluminescent structures", *Proceedings of SPIE 6192, Organic Optoelectronics and Photonics II*, pp. 619221–619221, 2006.

[CHA 11] CHAKAROUN M., COENS A., FABRE N. *et al.*, "Optimal design of a microcavity organic laser device under electrical pumping", *Optics Express*, vol. 19, pp. 493–505, 2011.

[CHA 13] CHAKAROUN M., ANTONY R., FISCHER A.P.A. *et al.*, "Enhanced electron injection and stability in organic light-emitting devices using an ion beam assisted cathode", *Solid State Sciences*, vol. 15, pp. 84–90, 2013.

[CHA 04] CHAN M.Y., LAI S.L., FUNG M.K. *et al.*, "Impact of the metal cathode and CsF buffer layer on the performance of organic light-emitting devices", *Journal of Applied Physics*, vol. 95, pp. 5397–5402, 2004.

[CHE 00] CHEN C.H., TANG C.W., SHI J. *et al.*, Green organic electroluminescent devices, US Patent US6020078 A, 2000.

[CHE 04] CHEN S.F., WANG C.W., "Influence of the hole injection layer on the luminescent performance of organic light-emitting diodes", *Applied Physics Letters*, vol. 85, pp. 765–767, 2004.

[CHE 07] CHEN Y., LI Z., ZHANG Z. *et al.*, "Nanoimprinted circular grating distributed feedback dye laser", *Applied Physics Letters*, vol. 91, pp. 051109-3, 2007.

[CHE 16] CHEN Y.H., LIN C.C., HUANG M.J. *et al.*, "Superior upconversion fluorescence dopants for highly efficient deep-blue electroluminescent devices", *Chemical Science*, vol. 7, pp. 4044–4051, 2016.

[CHR 08] CHRISTIANSEN M.B., KRISTENSEN A., XIAO S. *et al.*, "Photonic integration in k-space: enhancing the performance of photonic crystal dye lasers", *Applied Physics Letters*, vol. 93, pp. 231101–231104, 2008.

[CIA 11] CIAS P., SLUGOVC C., GESCHEIDT G., "Hole transport in triphenylamine based OLED devices: from theoretical modeling to properties prediction", *Journal of Physical Chemistry A*, vol. 115, pp. 14519–14525, 2011.

[CLA 07] CLARK J., SILVA C., FRIEND R.H. *et al.*, "Role of intermolecular coupling in the photophysics of disordered organic semiconductors: aggregate emission in regioregular polythiophene", *Physical Review Letters*, vol. 98, pp. 206406–206406-4, 2007.

[COE 13] COENS A., Diode électroluminescente organique en microcavité verticale à miroirs diélectriques multicouches, PhD Thesis, Paris 13 University, 2013.

[DAN 01] D'ANDRADE B.W., BALDO M., ADACHI C. *et al.*, "High-efficiency yellow double-doped organic light-emitting devices based on phosphor-sensitized fluorescence", *Applied Physics Letters*, vol. 79, pp. 1045–1047, 2001.

[DEE 00] DEEGAN R.D., BAKAJIN O., HUBER G. *et al.*, "Contact line deposits in an evaporating drop", *Physical Review E*, vol. 62, pp. 756–776, 2000.

[DEN 99] DENG Z.B., DING X.M., LEE S.T. *et al.*, "Enhanced brightness and efficiency in organic electroluminescent devices using SiO_2 buffer layer", *Thin Solid Films*, vol. 74, pp. 2227–2229, 1999.

[DEX 53] DEXTER D.L., "A theory of sensitized luminescence in solids", *The Journal of Chemical Physics*, vol. 21, pp. 836–850, 1953.

[DON 08] DONG Y., ZHAO H., SONG J. *et al.*, "Low threshold two-dimensional organic photonic crystal distributed feedback laser with hexagonal symmetry based on SiN", *Applied Physics Letters*, vol. 92, pp. 223309–223312, 2008.

[FAN 03] FANG J., MA D., "Efficient red organic light-emitting devices based on a europium complex", *Applied Physics Letters*, vol. 83, pp. 4041–4043, 2003.

[FÖR 48] FÖRSTER T., "Zwischenmolekulare energiewanderung und fluoreszenz", *Annalen der Physik*, vol. 6, pp. 55–75, 1948.

[FOW 28] FOWLER R.H., NORDHEIM L., "Electron emission in intense electric fields", *Proceedings of Royal Society London Series*, vol. 119, pp. 173–181, 1928.

[FRO 97] FROLOV S.V., GELLERMANN W., OZAKI M. *et al.*, "Cooperative emission in π-conjugated polymer thin films", *Physical Review Letters*, vol. 78, pp. 729–732, 1997.

[GAR 96] GARSTEIN Y.N., CONWELL E.M., "Field-dependent thermal injection into a disordered molecular insulator", *Chemical Physics Letters*, vol. 255, pp. 93–98, 1996.

[GÉR 98] GÉRARD J.M., SERMAGE B., GAYRAL B. *et al.*, "Enhanced spontaneous emission by quantum boxes in a monolithic optical microcavity", *Physical Review Letters*, vol. 81, pp. 1110–1113, 1998.

[GOU 12] GOURDON F., CHAKAROUN M., FABRE N. *et al.*, "Optically pumped lasing from organic two-dimensional planar photonic crystal microcavity", *Applied Physics Letters*, vol. 100, pp. 213304-4, 2012.

[GWI 09] GWINNER M.C., KHODABAKHSH S., SONG M.H. *et al.*, "Integration of a rib waveguide distributed feedback structure into a light-emitting polymer field-effect transistor", *Advanced Functional Materials*, vol. 19, pp. 1360–1370, 2009.

[HAR 05] HARBERS R., STRASSER P., CAIMI D. *et al.*, "Enhanced feedback in organic photonic-crystal lasers", *Applied Physics Letters*, vol. 87, pp. 151121–151124, 2005.

[HAY 95] HAYES G.R., SAMUEL I.D.W., PHILLIPS R.T., "Exciton dynamics in electroluminescent polymers studied by femtosecond time-resolved photoluminescence spectroscopy", *Physical Review B*, vol. 52, pp. 11569–11572, 1995.

[HEI 00] HEIKO J.U., SAWAT W.Z.M., FRANZ E. *et al.*, "Large-area single-mode selectively oxidized VCSELs: approaches and experimental", *Proceedings of SPIE 3946, Vertical-Cavity Surface-Emitting Lasers IV*, pp. 207–218, 2000.

[HEL 03] HELIOTIS G., XIA R., BRADLEY D.D.C. *et al.*, "Blue, surface-emitting, distributed feedback polyfluorene lasers", *Applied Physics Letters*, vol. 83, pp. 2118–2120, 2003.

[HEL 04] HELIOTIS G., XIA R., TURNBULL G.A. *et al.*, "Emission characteristics and performance comparison of polyfluorene lasers with one- and two-dimensional distributed feedback", *Advanced Functional Materials*, vol. 14, pp. 91–97, 2004.

[HID 96] HIDE F., SCHWARTZ B.J., GARCIA D.M.A. *et al.*, "Laser emission from solutions and films containing semiconducting polymer and titanium dioxide nanocrystals", *Chemical Physics Letters*, vol. 256, pp. 424–430, 1996.

[HOS 95] HOSOKAWA C., HIGASHI H., NAKAMURA H. *et al.*, "Highly efficient blue electroluminescence from a distyrylene emitting layer with a new dopant", *Applied Physics Letters*, vol. 67, pp. 3853–3855, 1995.

[HUG 05] HUGHES G., BRYCE M.R., "Electron-transporting materials for organic electroluminescent and electrophosphorescent devices", *Journal of Materials Chemistry*, vol. 15, pp. 94–107, 2005.

[HWA 05] HWANG S.W., OH H.S., KANG S.J., "A study of the fabrication and the characteristics of an organic light-emitting device using BCP", *Journal of the Korean Physical Society*, vol. 47, pp. 34–36, 2005.

[ISH 99] ISHII H., SUGIYAMA K., ITO E. *et al.*, "Energy level alignment and interfacial electronic structures at organic metal and organic organic interfaces", *Advanced Materials*, vol. 11, pp. 605–625, 1999.

[JAB 98] JABBOUR G.E., KIPPELEN B., ARMSTRONG N.R. *et al.*, "Aluminum based cathode structure for enhanced electron injection in electroluminescent organic devices", *Applied Physics Letters*, vol. 73, pp. 1185–1187, 1998.

[JAB 00] JABBOUR G.E., KIPPELEN B., ARMSTRONG N.R. *et al.*, "Aluminium based cathode structure for enhanced electron injection layer", *Journal of Luminescence*, vol. 87, pp. 1185–1187, 2000.

[JEA 02] JEAN F., MULOT J.Y., GEFFROY B. *et al.*, "Microcavity organic light emitting diodes on silicon", *Applied Physics Letters*, vol. 81, pp. 1717–1719, 2002.

[JIA 00] JIANG H., ZHOU Y., OOI B.S. *et al.*, "Improvement of organic light emitting diodes performance by the insertion of a Si_3N_4 layer", *Thin Solid Films*, vol. 363, pp. 25–28, 2000.

[JOR 06] JORDAN G., FLÄMMICH M., RÜTHER M. *et al.*, "Light amplification at 501 nm and large nanosecond optical gain in organic dye-doped polymeric waveguides", *Applied Physics Letters*, vol. 88, pp. 161114–161117, 2006.

[JUN 01] JUNG B.Y., KIM N.Y., LEE C.H. *et al.*, "Optical properties of Fabry–Pérot microcavity with organic light emitting materials", *Current Applied Physics*, vol. 1, pp. 175–181, 2001.

[KAK 02] KAKO S., SOMEYA T., ARAKAWA Y., "Observation of enhanced spontaneous emission coupling factor in nitride-based vertical-cavity surface-emitting laser", *Applied Physics Letters*, vol. 80, pp. 722–724, 2002.

[KAL 96] KALINOWSKI J., "Organic electroluminescence: materials and devices", *Proc. SPIE 2780, Metal/Nonmetal Microsystems: Physics, Technology, and Applications*, pp. 293–303, 1996.

[KAM 06] KAMPEN T.U., "Electronic structure of organic interfaces – a case study on perylene derivatives", *Applied Physics A*, vol. 3, pp. 457–470, 2006.

[KAN 97] KANAI H., ICHINOSAWA S., SATO Y., "Effect of aromatic diamines as a cathode interface layer", *Synthetic Metals*, vol. 91, pp. 195–196, 1997.

[KAN 04] KAN Y., WANG L., DUAN L. *et al.*, "Highly-efficient blue electroluminescence based on two emitter isomers", *Applied Physics Letters*, vol. 84, pp. 1513–1515, 2004.

[KAN 07] KANG J.W., LEE S.H., PARK H.D. *et al.*, "Low roll-off of efficiency at high current density in phosphorescent organic light emitting diodes", *Applied Physics Letters*, vol. 90, pp. 223508–223508-3, 2007.

[KAR 93] KARASAWA T., MIYATA Y., "Electrical and optical properties of indium tin oxide thin films deposited on unheated substrate by D.C. reactive sputtering", *Thin Solid Films*, vol. 223, pp. 135–139, 1993.

[KAR 07] KARNUTSCH C., PFLUMM C., HELIOTIS G. *et al.*, "Improved organic semiconductor lasers based on a mixed-order distributed feedback resonator design", *Applied Physics Letters*, vol. 90, pp. 131104–131104-3, 2007.

[KAW 05] KAWAMURA Y., GOUSHI K., BROOKS J. et al., "100% phosphorescence quantum efficiency of Ir(III) complexes in organic semiconductor films", *Applied Physics Letters*, vol. 86, pp. 071104–071104-3, 2005.

[KAZ 99] KAZUHIKO S., HIROSHI O., EISUKE I. et al., "Energy level alignment and band bending at organic interfaces", *Proc. SPIE 3797, Organic Light-Emitting Materials and Devices III*, pp. 178–188, 1999.

[KES 13] KESSLER F., WATANABE Y., SASABE H. et al., "High-performance pure blue phosphorescent OLED using a novel bis-heteroleptic iridium(III) complex with fluorinated bipyridyl ligands", *Journal of Materials Chemistry C*, vol. 1, pp. 1070–1075, 2013.

[KHA 16] KHADIR S., Effets des résonances plasmon de surface localisé sur les performances optiques et électriques des diodes électroluminescentes organiques, PhD Thesis, Tizi-Ouzou University, Algeria, 2016.

[KIM 07] KIM H.K., BYUN Y.H., DAS R.R. et al., "Small molecule based and solution processed highly efficient red electrophosphorescent organic light emitting devices", *Applied Physics Letters*, vol. 91, pp. 093512093512-3, 2007.

[KIT 04] KITAMURA M., IMADA T., KAKO S. et al., "Time-of-flight measurement of lateral carrier mobility in organic thin films", *Japanese Journal of Applied Physics*, vol. 43, pp. 2326–2329, 2004.

[KIT 05] KITAMURA M., IWAMOTO S., ARAKAWA Y., "Enhanced light emission from an organic photonic crystal with a nanocavity", *Applied Physics Letters*, vol. 87, pp. 151119–151119-3, 2005.

[KLE 12] KLEEMANN H., LÜSSEM B., LEO K., "Controlled formation of charge depletion zones by molecular doping in organic pindiodes and its description by the Mott-Schottky relation", *Journal of Applied Physics*, vol. 111, pp. 123722–123722-7, 2012.

[KOC 07] KOCH N., "Organic electronic devices and their functional interfaces", *Chemphyschemstry*, vol. 8, no. 10, pp. 1438–1455, 2007.

[KÖG 71] KÖGELNIK H., SHANK C.V., Coupled-Wave Theory of Distributed Feedback Lasers, Bell Telephone Laboratories, Holmdel, 1971.

[KÖG 72] KÖGELNIK H., SHANK C.V., "Coupled-wave theory of distributed feedback lasers", *Journal of Applied Physics*, vol. 43, pp. 2327–2335, 1972.

[KOS 05] KOSCHORRECK M., GEHLHAAR R., LYSSENKO V.G. et al., "Dynamics of a high-Q vertical-cavity organic laser", *Applied Physics Letters*, vol. 87, pp. 181108–181111, 2005.

[KOZ 98] KOZLOV V.G., BULOVIC V., BURROWS P.E. *et al.*, "Study of lasing action based on Förster energy transfer in optically pumped organic semiconductor thin films", *Journal of Applied Physics*, vol. 84, pp. 4096–4109, 1998.

[KOZ 00] KOZLOV V.G., PARTHASARATHY G., BURROWS P.E. *et al.*, "Structures for organic diode lasers and optical properties of organic semiconductors under intense optical and electrical excitations", *IEEE Journal of Quantum Electronics*, vol. 36, pp. 18–26, 2000.

[KUG 97] KUGLER T., LÖGDLUND M., SALANECK W.R., "Electronic and chemical structure of conjugated polymer surfaces and interfaces: application in polymer-based light emitting devices", unpublished, *4th European Conference on Molecular Electronics*, Cambridge, 1997.

[KUM 12] KUMAR S., HU Q., "Investigation of possible microcavity effect on lasing threshold of nonradiative-scattering-dominated semiconductor lasers", *Applied Physics Letters,* vol. 100, pp. 041105-4, 2012.

[LAT 06] LATTANTE S., ROMANO F., CARICATO A.P. *et al.*, "Low electrode induced optical losses in organic active single layer polyfluorene waveguides with two indium tin oxide electrodes deposited by pulsed laser deposition", *Applied Physics Letters,* vol. 89, pp. 031108–031111, 2006.

[LEE 04a] LEE M.T., CHEN H.H., LIAO C.H. *et al.*, "Stable styrylamine-doped blue organic electroluminescent device based on 2-methyl-9,10-di (2-naphthyl) anthracene", *Applied Physics Letters*, vol. 85, pp. 3301–3303, 2004.

[LEE 04b] LEE M.T., CHEN H.H., TSAI C.H. *et al.*, "Development of highly efficient and stable OLEDs", *Proc. Int. Meeting Information Display*, Daegu, Korea, pp. 265–268, 2004.

[LEE 05] LEE J.H., WU C.I., LIU S.W. *et al.*, "Mixed host organic light-emitting devices with low driving voltage and long lifetime", *Applied Physics Letters*, vol. 86, pp. 103506–103506-3, 2005.

[LEE 15] LEE H., HWANG Y., WON T., "Effect of inserting a hole injection layer in organic light-emitting diodes: a numerical approach", *Journal of the Korean Physical Society*, vol. 66, pp. 100–103, 2015.

[LIS 01] LIST E.J.W., KIM C.H., NAIK A.K. *et al.*, "Interaction of singlet excitons with polarons in wide band-gap organic semiconductors: a quantitative study", *Physical Review B*, vol. 64, pp. 155204–155215, 2001.

[LIU 09] LIU X., LI H., SONG C. *et al.*, "Microcavity organic laser device under electrical pumping", *Optics Letters*, vol. 34, pp. 503–505, 2009.

[LU 04] LU W., ZHONG B., Ma D., "Amplified spontaneous emission and gain from optically pumped films of dye-doped polymers", *Applied Optics*, vol. 43, pp. 5074–5078, 2004.

[MAI 60] MAIMAN T.H., "Stimulated optical radiation in ruby", *Nature*, vol. 187, pp. 493–494, 1960.

[MAL 99] MALLIARAS G.G., SCOTT J.C., "The roles of injection and mobility in organic light emitting diodes", *Journal of Applied Physics*, vol. 83, pp. 5399–5403, 1999.

[MAR 05] MARTIRADONNA L., De VITTORIO M., TROISI L. *et al.*, "Fabrication of hybrid organic-inorganic vertical microcavities through imprint technology", *Microelectronic Engineering*, vol. 78, pp. 593–597, 2005.

[MAS 00] MASENELLI B., CALLARD S., GAGNAIRE A. *et al.*, "Fabrication and characterization of organic semiconductor-based microcavities", *Thin Solid Films*, vol. 364, pp. 264–268, 2000.

[MAS 01] MASON M.G., TANG C.W., HUNG L.S. *et al.*, "Interfacial chemistry of Alq and LiF with reactive metals", *Journal of Applied Physics*, vol. 89, pp. 4986–7992, 2001.

[MCG 98] MCGEHEE M.D., DÍAZ-GARCIA M.A., HIDE F. *et al.*, "Semiconducting polymer distributed feedback lasers", *Applied Physics Letters*, vol. 72, pp. 1536–1538, 1998.

[MCG 00] MCGEHEE M.D., HEGEER A.J., "Semiconducting (conjugated) polymers as materials for solid-state lasers", *Advanced Materials*, vol. 12, pp. 1655–1668, 2000.

[MOL 03] MOLITON A., *Optoélectronique moléculaire et polymère: des concepts aux composants*, Springer, Paris, 2003.

[MOL 11] MOLITON A., *Électronique et optoélectronique organiques*, Springer, 2011.

[MOS 92] MOSES D., "High quantum efficiency luminescence from a conducting polymer in solution: a novel polymer laser dye", *Applied Physics Letters*, vol. 60, pp. 3215–3218, 1992.

[MUR 10] MURSHIDY M.M., ADAWI A.M., FRY P.W. *et al.*, "The optical properties of hybrid organic–inorganic L3 nanocavities", *Journal of the Optical Society of America B*, vol. 27, no. 2, pp. 215–221, 2010.

[NAK 05] NAKANOTANI H., SASABE H., ADACHIA C., "Singlet-singlet and singlet-heat annihilations in fluorescence-based organic light-emitting diodes under steadystate high current density", *Applied Physics Letters*, vol. 86, pp. 213506–213509, 2005.

[NAR 13] NARAYAN K., VARADHARAJAPERUMAL S., MOHAN R.G. *et al.*, "Effect of thickness variation of hole injection and hole blocking layers on the performance of fluorescent green organic light emitting diodes", *Current Applied Physics*, vol. 13, pp. 18–25, 2013.

[NAY 10] NAYAK P.K., AGARWAL N., ALI F. *et al.*, "Blue and white light electroluminescence in a multilayer OLED using a new aluminum complex", *Journal of Chemical Sciences*, vol. 122, pp. 847–855, 2010.

[NIN 13] NING C.Z., "What is laser threshold", *IEEE of Selected Topics in Quantum Electronics*, vol. 19, pp. 1503604-1503604, 2013.

[PAL 68] PALMER D.A., "Standard observer for large-field photometry at any level", *Journal of the Optical Society of America*, vol. 58, pp. 1296–1299, 1968.

[PAR 09] PARK J.J., PARK T.J., JEON W.S. *et al.*, "Small molecule interlayer for solution processed phosphorescentorganic light emitting device", *Organic Electronics*, vol. 10, pp. 189–193, 2009.

[PAR 01] PARTHASARATHY G., SHEN C., KAHN A. *et al.*, "Lithium doping of semiconducting organic charge transport materials", *Journal of Applied Physics*, vol. 89, pp. 4986–4992, 2001.

[PEI 55] PEIERLS R., *Quantum Theory of Solids*, Oxford University Press, Oxford, 1955.

[PEL 02] PELTON M., VUCKOVIC J., GLENN S.S. *et al.*, "Three-dimensionally confined modes in micropost microcavities: quality factors and Purcell factors", *IEEE Journal of Quantum Electronics*, vol. 38, no. 2, pp. 170–177, 2002.

[PIR 00] PIROMREUN P., OH H., SHEN Y. *et al.*, "Role of CsF on electron injection into a conjuguated polymer", *Applied Physics Letters,* vol. 77, pp. 2403–2405, 2000.

[QIU 06] QIU X.J., TAN X.W., WANG Z. *et al.*, "Tunable, narrow, and enhanced electroluminescent emission from porous-silicon-reflector-based organic microcavities", *Journal of Applied Physics*, vol. 100, pp. 074503–74501-6, 2006.

[RAB 09] RABBANI H.H., FORGET S., CHÉNAIS S. *et al.*, "Laser operation in nondoped thin films made of small-molecule organic red emitter", *Applied Physics Letters*, vol. 95, pp. 033305–033305-3, 2009.

[RAB 05] RABE T., HOPING M., SCHNEIDER D. *et al.*, "Threshold reduction in polymer lasers based on poly(9,9-dioctylfluorene) with statistical Binaphthyl units", *Advanced Functional Materials,* vol. 15, pp. 1188–1192, 2005.

[RAM 06] RAMMAL W., Réalisation de diodes électroluminescentes souples et caractérisations, PhD Thesis, University of Limoges, 2006.

[REU 04] REUFER M., RIECHEL S., LUPTON J.M. *et al.*, "Low threshold polymeric distributed feedback lasers with metallic contacts", *Applied Physics Letters*, vol. 84, pp. 3262–3264, 2004.

[REU 05] REUFER M., WALTER M., LAGOUDAKIS P. *et al.*, "Spin-conserving carrier recombination in conjugated polymers", *Nature Materials*, vol. 4, pp. 1–7, 2005.

[RIC 01] RICHARDSON O.W., *On the Negative Radiation from Hot Platinum*, Cambridge Philosophical Society, 1901.

[RIE 00] RIECHEL S., LEMMER U., FELDMANN J. *et al.*, "Laser modes in organic solid-state distributed feedback lasers", *Applied Physics B*, vol. 71, no. 6, pp. 897–900, 2000.

[RIE 01] RIECHEL S., LEMMER U., FELDMANN J. *et al.*, "Very compact tunable solid-state laser utilizing a thin-film organic semiconductor", *Optics Letters*, vol. 26, pp. 593–595, 2001.

[RIE 06] RIEDL T., RABE T., JOHANNES H.H. *et al.*, "Tunable organic thin-film laser pumped by an inorganic violet diode laser", *Applied Physics Letters*, vol. 88, pp. 241116–241119, 2006.

[RIE 15] RIEDEL D., DLUGOSCH J., WEHLUS T. *et al.*, "Extracting and shaping the light of OLED devices", *Proceedings of the SPIE, Organic Light Emitting Materials and Devices XIX*, vol. 9566, pp. 95661H1–95661H,9, 2015.

[RUT 13] RUTLEDGE S.A., HELMY A.S., "Carrier mobility enhancement in poly(3,4-ethylenedioxythiophene) – poly(styrenesulfonate) having undergone rapid thermal annealing", *Journal of Applied Physics*, vol. 114, pp. 133708–133708-5, 2013.

[SAK 08] SAKATA H., TAKEUCHI H., "Diode-pumped polymeric dye lasers operating at a pump power level of 10 mW", *Applied Physics Letters*, vol. 92, pp. 113310–113310-3, 2008.

[SAM 07] SAMUEL I.D.W., TURNBULL G.A., "Organic semiconductor lasers", *Chemical Reviews*, vol. 107, pp. 1272–1295, 2007.

[SAM 09] SAMUEL I.D.W., NAMDAS E.B., TURNBULL G.A., "How to recognize lasing", *Nature Photonics*, vol. 3, pp. 546–549, 2009.

[SCH 07] SCHANDA J., *Colorimetry: Understanding the CIE System*, John Wiley & Sons, 2007.

[SCH 01] SCHERFA U., RIECHELB S., LEMMERB U. *et al.*, "Conjugated polymers: lasing and stimulated emission", *Current Opinion in Solid State and Materials Science*, vol. 5, no. 2, pp. 143–154, 2001.

[SCH 98] SCHWEITZER B., WEGMANN G., GIESSEN H. *et al.*, "The optical gain mechanism in solid conjugated polymers", *Applied Physics Letters*, vol. 72, pp. 2933–2936, 1998.

[SEI 07] SEIJI F., KAZUKI B., YASUAKI M. *et al.*, "Laser oscillations of whispering gallery modes in thiophene/phenylene co-oligomer microrings", *Applied Physics Letters*, vol. 91, pp. 021104,1–021104,3, 2007.

[SHA 71] SHAKLEE K.L., LEHENY R.F., "Direct determination of optical gain in semiconductor crystals", *Applied Physics Letters*, vol. 18, pp. 475–447, 1971.

[SHI 02a] SHIA J., TANG C.W., "Anthracene derivatives for stable blue-emitting organic electroluminescence devices", *Applied Physics Letters*, vol. 80, pp. 3201–3203, 2002.

[SHI 02b] SHIH H.T., LIN C.H., SHIH H.H. *et al.*, "High-performance blue electroluminescent devices based on a biaryl", *Advanced Materials*, vol. 14, pp. 1409–1412, 2002.

[SHI 97] SHIROT Y., KUWABARA Y., OKUD D. *et al.*, "Starburst molecules based on π-electron systems as materials for organic electroluminescent devices", *Journal of Luminescence*, vol. 72, pp. 985–991, 1997.

[SHO 65] SHORT G.D., HERCULES D.M., "Electroluminescence of organic compounds. the role of gaseous discharge in the excitation process", *Journal of the American Chemical Society*, vol. 87, pp. 1439–1442, 1965.

[SIV 09] SIVASUBRAMANIAM V., BRODKORB F., HANNING S. *et al.*, "Degradation of HTL layers during device operation in PhOLEDs", *Solid State Sciences*, vol. 11, pp. 1933–1940, 2009.

[SOK 96] SOKOLIK I., PRIESTLEY R., WALSER A. *et al.*, "Bimolecular reactions of singlet excitons in tris(8-hydroxyquinoline) aluminum", *Applied Physics Letters*, vol. 69, pp. 4168–4171, 1996.

[SOR 66] SOROKIN P., LANKARD J.R., "Stimulated emission observed from an organic dye, chloro-aluminum phthalocyanine", *IBM Journal of Research and Development.*, vol. 10, pp. 162–163, 1966.

[SPE 05] SPEHR T., SIEBERT A., LIEKER F.T. *et al.*, "Organic solid-state ultraviolet-laser based on spiro-terphenyl", *Applied Physics Letters*, vol. 87, pp. 161103–161106, 2005.

[STA 99] STAUDIGEL J., STÖSSEL M., STEUBER F. *et al.*, "Comparison of mobility and hole current activation energy in the space charge trap-limited regime in a starburst amine", *Applied Physics Letters*, vol. 75, pp. 217–219, 1999.

[STÖ 99] STÖBEL M., STAUDIGEL J., STEUBER F. *et al.*, "Space-charge-limited electron currents in 8-hydroxyquinoline aluminum", *Applied Physics Letters*, vol. 76, pp. 15–117, 1999.

[STO 08] STOCKMAN M.I., "Spasers explained", *Nature Photonics*, vol. 2, pp. 327–329, 2008.

[STO 09] STOCKMAN M.I., BERGMAN D.J., Surface plasmon amplification by stimulated emission of radiation (spaser), US Patent 7,569,188, 2009.

[STO 10] STOCKMAN M.I., "The spaser as a nanoscale quantum generator and ultrafast amplifier", *Journal of Optics*, vol. 12, pp. 024004–024004-13, 2010.

[STO 11a] STOCKMAN M.I., "Spaser action, loss compensation, and stability in plasmonic systems with gain", *Physical Review Letters*, vol. 106, pp. 156802–156802-4, 2011.

[STO 11b] STOCKMAN M.I., "Loss compensation by gain and spasing", *Philosophical Transactions of the Royal Society A*, vol. 369, pp. 3510–3524, 2011.

[STR 11] STRAUD S., JAHNKE F., "Single quantum dot nanolaser", *Laser Photonics Reviews*, vol. 5, pp. 607–633, 2011.

[SUN 06] SUN Y., GIEBINK N.C., KANNO H. *et al.*, "Management of singlet and triplet excitons for efficient white organic light-emitting devices", *Nature*, vol. 440, pp. 908–912, 2006.

[SUN 08] SUN G., KHURGIN J.B., SOREF R.A., "Plasmonic lightemission enhancement with isolated metal nanoparticles and their coupled arrays", *Journal of the Optical Society of America B*, vol. 25, pp. 1748–1755, 2008.

[SUN 09] SUN G., KHURGIN G.B., SOREF R.A., "Practical enhancement of photoluminescence by metal nanoparticles", *Applied Physics Letters*, vol. 94, pp. 101103–101103-06, 2009.

[SUN 11a] SUN G., KHURGIN G.B., "Plasmon enhancement of luminescence by metal nanoparticles", *IEEE Journal of Selected Topics in Quantum Electronics*, vol. 17, pp. 110–118, 2011.

[SUN 11b] SUN G., KHURGIN J.B., "Theory of optical emission enhancement by coupled metal nanoparticles: an analytical approach", *Applied Physics Letters*, vol. 98, 113116–113119, 2011.

[SUZ 04] SUZUKI K., SENO A., TANABE H. *et al.*, "New host materials for blue emitters", *Synthetic Metals*, vol. 143, pp. 89–96, 2004.

[TAK 08] TAKENOBU T., BISRI S.Z., TAKAHASHI T. *et al.*, "High current density in light-emitting transistors of organic single crystals", *Physical Review Letters*, vol. 100, pp. 066601–066605, 2008.

[TAN 08] TANAKA I., TOKITO S., "Energy-transfer processes between phosphorescent guest and fluorescent host molecules in phosphorescent OLEDs", in YERSIN H. (ed.), *Highly Efficient OLEDs with Phosphorescent Materials*, Wiley–VCH Verlag GmbH & Co. KGaA, Weinheim, 2008.

[TAN 87] TANG C.W., VANSLYKE S.A., "Organic electroluminescent diodes", *Applied Physics Letters*, vol. 51, pp. 913–915, 1987.

[TES 96] TESSLER N., DENTON G.J., FRIEND R.H., "Lasing from conjugated-polymer microcavities", *Nature,* vol. 382, pp. 695–697, 1996.

[TES 98] TESSLER N., HARRISON N.T., FRIEND R.H., "High peak brightness polymer light-emitting diodes", *Advanced Materials*, vol. 10, pp. 64–68, 1998.

[TES 99] TESSLER N., "Lasers based on semiconducting organic materials", *Advanced Materials*, vol. 11, pp. 363–370, 1999.

[TOK 99] TOKITO S., TSUTSUI T., TAGA Y., "Microcavity organic light-emitting diodes for strongly directed pure red, green, and blue emissions", *Journal of Applied Physics*, vol. 86, pp. 2407–2411, 1999.

[TOK 03] TOKITO S., IIJIMA T., SUZURI Y. *et al.*, "Confinement of triplet energy on phosphorescent molecules for highly-efficient organic blue-light-emitting devices", *Applied Physics Letters*, vol. 83, pp. 569–571, 2003.

[TSU 07] TSUBOI T., MURAYAMA H., YEH S.J. *et al.*, "Energy transfer between organic fluorescent CBP host and blue phosphorescent FIrpic and FIrN4 guests", *Optical Materials*, vol. 29, pp. 1299–1304, 2007.

[TSU 04] TSUBOYAMA A., IWAWAKI H., FURUGORI M. *et al.*, "Homoleptic cyclometalated iridium complexes with highly efficient red phosphorescence and application to organic light-emitting diode", *Journal of the American Chemical Society*, vol. 125, pp. 12971–12979, 2004.

[VAL 12] VALEUR B., *Molecular Fluorescence: Principles and Applications Molecular Fluorescence: Principles and Applications*, John Wiley & Sons, 2012.

[VAS 05] VASDEKIS A.E., TURNBULL G.A., SAMUEL I.D.W. *et al.*, "Low threshold edge emitting polymer distributed feedback laser based on a square lattice", *Applied Physics Letters*, vol. 86, pp. 161102–161105, 2005.

[VAS 06] VASDEKIS A.E., TSIMINIS G., RIBIERRE J.C. *et al.*, "Diode pumped distributed Bragg reflector lasers based on a dye-to-polymer energy transfer blend", *Optics Express*, vol. 14, pp. 9211–9216, 2006.

[WEN 05] WEN S.W., LEE M.T., CHEN C.H., "Recent development of blue fluorescent OLED materials and devices", *Journal of Display Technology*, vol. 1, pp. 90–99, 2005.

[WET 11] WETZELAER G.A.H., KUIK M., NICOLAI H.T. *et al.*, "Trap-assisted and Langevin-type recombination in organic light-emitting diodes", *Physical Review B*, vol. 83, pp. 165204–165204-5, 2011.

[WOL 99] WOLF U., ARKHIPOV I.V., BASSLER H., "Current injection from a metal to a disordered hopping system, I. Monte Carlo simulation", *Physical Review B*, vol. 59, pp. 7507–5713, 1999.

[WU 97] WU C.C., WU C.I., STURM J.C. *et al.*, "Surface modification of indium tin oxide by plasma treatment: an effective method to improve the efficiency, brightness, and reliability of organic light emitting devices", *Applied Physics Letters*, vol. 70, pp. 1348–1350, 1997.

[XU 06] XU X.J., YU G., "Electrode modification in organic light-emitting diodes", *Displays,* vol. 27, no. 1, pp. 24–34, 2006.

[YAN 08] YANG Y., TURNBULL G.A., SAMUEL I.D.W., "Hybrid optoelectronics: a polymer laser pumped by a nitride light-emitting diode", *Applied Physics Letters*, vol. 92, pp. 163306–163309, 2008.

[YAN 09] YANG K.Y., CHOI K.C., AHN C.W., "Surface plasmon-enhanced spontaneous emission rate in an organic light-emitting device structure: cathode structure for plasmonic application", *Applied Physics Letters*, vol. 94, pp. 173301–173305, 2009.

[YAN 12] YANG Q., HAO Y., WANG Z. *et al.*, "Double-emission-layer green phosphorescent OLED based on LiF-doped TPBi as electron transport layer for improving efficiency and operational lifetime", *Synthetic Metals*, vol. 162, pp. 398–401, 2012.

[YU 02] YU M.X., DUAN J.D., LIN C.H. *et al.*, "Diaminoanthracene derivatives as high-performance green host electroluminescent materials", *Chemistry of Materials*, vol. 14, pp. 3958–3963, 2002.

Index

Printed in the United States
By Bookmasters